Assessing Revolutionary and Insurgent Strategies

CASE STUDIES IN INSURGENCY AND REVOLUTIONARY WARFARE— FOSTERING EFFECTIVE COUNTER UNCONVENTIONAL WARFARE/OCCUPATION

United States Army Special Operations Command

Case Studies in Insurgency and Revolutionary Warfare—Fostering Effective Counter Unconventional Warfare/Occupation is a work of the United States Government in accordance with Title 17, United States Code, sections 101 and 105.

Published by Conflict Research Group.

First published by USASOC in 2019

CONFLICT RESEARCH GROUP

Cite me as:

Leonhard, Robert, et al. *Case Studies in Insurgency and Revolutionary Warfare—Fostering Effective Counter Unconventional Warfare/Occupation*. Fort Bragg: US Army Special Operations Command, 2019.

Assessing Revolutionary and Insurgent Strategies

CASE STUDIES IN INSURGENCY AND REVOLUTIONARY WARFARE— FOSTERING EFFECTIVE COUNTER UNCONVENTIONAL WARFARE/OCCUPATION

Robert Leonhard, Author

Johns Hopkins University Applied Physics Laboratory
(JHU/APL)

United States Army Special Operations Command

ASSESSING REVOLUTIONARY AND INSURGENT STRATEGIES

The Assessing Revolutionary and Insurgent Strategies (ARIS) series consists of a set of case studies and research conducted for the US Army Special Operations Command by the National Security Analysis Department of the Johns Hopkins University Applied Physics Laboratory.

The purpose of the ARIS series is to produce a collection of academically rigorous yet operationally relevant research materials to develop and illustrate a common understanding of insurgency and revolution. This research, intended to form a bedrock body of knowledge for members of the Special Forces, will allow users to distill vast amounts of material from a wide array of campaigns and extract relevant lessons, thereby enabling the development of future doctrine, professional education, and training.

From its inception, ARIS has been focused on exploring historical and current revolutions and insurgencies for the purpose of identifying emerging trends in operational designs and patterns. ARIS encompasses research and studies on the general characteristics of revolutionary movements and insurgencies and examines unique adaptations by specific organizations or groups to overcome various environmental and contextual challenges.

The ARIS series follows in the tradition of research conducted by the Special Operations Research Office (SORO) of American University in the 1950s and 1960s, by adding new research to that body of work and in several instances releasing updated editions of original SORO studies.

RECENT VOLUMES IN THE ARIS SERIES

Casebook on Insurgency and Revolutionary Warfare: Volume I: 1927-1962 (2013)

Casebook on Insurgency and Revolutionary Warfare: Volume II: 1962-2009 (2012)

Human Factors Considerations of Undergrounds in Insurgencies (2013)

Undergrounds in Insurgent, Revolutionary, and Resistance Warfare (2013)

Understanding States of Resistance (2019)

Legal Implications of the Status of Persons in Resistance (2015)

Threshold of Violence (2019)

"Little Green Men": A Primer on Modern Russian Unconventional Warfare, Ukraine 2013-2014 (2015)

Science of Resistance (forthcoming)

TABLE OF CONTENTS

CHAPTER 1. INTRODUCTION AND SUMMARY 1

  Introduction... 2

    Methodology 2

    Terms 3

    The Nature of the Threat 3

    Non-military actions 6

    Military actions 9

CHAPTER 2. THE THREAT OF RUSSIAN MILITARY
OCCUPATION – HISTORICAL CASE STUDIES 12

  Introduction... 13

    Soviet and German Occupation, 1939-1991 14

  Other Cases of Russian Aggression During and After the Cold War... 16

    Lithuania, 1991 16

    Transnistria, 1990-92 17

    Serbian Krajina, 1991-95 18

    Chechnya, 1994-96 18

    Dagestan and the Second Chechen War, 1999-2009 19

    Georgia, 2008... 19

    The Color Revolutions... 21

CHAPTER 3. GUERRILLA MOVEMENTS AND LATVIA'S
RESISTANCE AGAINST THE SOVIETS AND RUSSIAN
FEDERATION 23

    Latvia's Quest for Independence, 1917-1920.................... 24

    Latvia's Democratic Experiment, 1918-1934................... 28

    The End of Democracy, 1934-39 31

      First Soviet Occupation, 1939-41 33

      Latvian Resistance 37

      German Invasion and Occupation, 1941-44 38

      Latvian Resistance to the Nazi Occupation 40

      Strategic Miscalculation.. 42

Soviet Reoccupation, 1944-56 44

The Latvian Resistance 45

CHAPTER 4. POPULIST MOVEMENTS AND POLAND'S
RESISTANCE AGAINST THE SOVIETS AND COMMUNIST
REGIME 50

Introduction 51

Polish Populist Uprisings—the KOR and Solidarity 53

The Worker's Defense Committee (KOR) 56

The Solidarity Movement 59

Karol Wojtyla 61

Lech Walesa 62

Solidarnosc 62

Conclusion 64

CHAPTER 5. OTHER CASES OF RUSSIAN AGGRESSION
DURING AND AFTER THE COLD WAR 67

Lithuania, 1991 68

Transnistria, 1990-92 68

Serbian Krajina, 1991-95 69

Chechnya, 1994-96 69

Dagestan and the Second Chechen War, 1999-2009 70

Georgia, 2008 71

The Color Revolutions 72

CHAPTER 6. OPERATION "GLADIO" AND THE RISK OF STAY-
BEHIND NETWORKS 75

CHAPTER 7. CONCLUSION 79

Conclusion and Findings 80

Potential Russian Aggression 81

Fostering Effective Counter-UW 84

APPENDIX 87

THE REPUBLIC OF NORTHARIA 88

Description 88

Fostering Effective Defense in the Republic of Northaria 88

BIBLIOGRAPHY 133

LIST OF ILLUSTRATIONS

Figure 1: Soviet Occupation of Eastern Europe, 1939 15

Figure 2: Karlis Ulmanis After the Coup of 1934 31

Figure 3: Soviet Invasion and Occupation, June 1940 35

Figure A-4: Protests over Language Discrimination[3] 114

Figure A-5: Soldiers of Odin Participating in Anti-Migrant Protest[4] 117

Figure A-6: Fence Separating the Northarian and Russian Border to Prevent Entrance of Refugees[5] 118

Figure A-7: Commemoration of Legionnaires as Veterans Lay Flowers at the Foot of the Liberation Statue 118

Figure A-8: Typical Unpaved Rural Road in Northaria. 122

Figure A-9: Diagram of Northaria's Electric Grid Infrastructure 124

LIST OF TABLES

Table A-1: Parties in the Majvendi and the European Parliament[1] 94

CHAPTER 1.
INTRODUCTION AND SUMMARY

INTRODUCTION

The objective of this study is to explore how to effectively prepare for and oppose unconventional warfare (UW) and/or military occupation in order to inform strategy formulation within vulnerable countries. The focus is on actual and potential aggression by the Russian Federation in Europe, and especially in Eastern Europe. The study looks at what measures a country can take to identify and resist both military and nonmilitary aggression. The latter can take many forms—from televised propaganda to coercive economic policies to the use of organized crime to the funding of political parties. With regard to military occupation, the study examines what actions a potential target state could take in the areas of logistics, communications, command, organization, intelligence, sabotage, subversion, and guerrilla operations so that if an aggressor invades and occupies the country, the population can begin to resist immediately.

The objective of this effort is therefore to examine two different but related problems. The first problem looks at resisting UW-- best demonstrated by Russia's campaign in Ukraine, 2013-present. Russian operations there featured the use of non-kinetic as well as kinetic operations to coerce the Kyiv government, foster and organize resistance, and annex Crimea. This type of aggression can include military operations, including incursions, raids, attacks, and defenses, etc., i.e., military operations short of a full-scale, deliberate invasion. The second problem is that of outright military occupation, such as occurred throughout Eastern Europe before, during, and after World War Two.

Methodology

This study will begin by examining the extensive historical record of resistance against Soviet and Russian aggression from World War Two through the present. The goal of this first part is not to restate the history, but to draw from it to derive key insights and lessons learned. The second part of the study then takes those lessons and applies them to a fictional East European country named "The Republic of Northaria." The authors chose to use a notional country in order to avoid fixation on one particular country and instead demonstrate pragmatic steps that any country could take to reduce its vulnerability.

The sources used come from a wide spectrum of articles, case studies, and books. The authors have taken care not to overestimate the threat or to fall victim to polemic. Root causes of conflict run deep, and what may appear as nefarious plans hatched in the Kremlin can in fact derive from a multitude of factors. Nevertheless, Russian aggression is real, and it presents an enduring problem—indeed, for some countries, the most significant security problem—in modern Europe.

When discussing the Republic of Northaria, the goal will be to describe the optimal preparation that the country's government and people could take to deter and resist Russian aggression. The intent is that real-world strategists can use the notional case of Northaria as a basis for strategic formulation in vulnerable countries.

Terms

The term "unconventional warfare" is used in this study in a general sense to describe irregular warfare (specifically Russia's New Generation Warfare), including political, diplomatic, military, economic, financial, cultural, social, religious, cyber, and information warfare.

The Nature of the Threat

This study considers how European countries can best prepare themselves to successfully defend against Russian aggression—either the hybrid warfare recently called "New Generation Warfare," or outright military invasion and occupation.[1] This presupposes that Russia indeed intends to wage such aggression, so it is logical to begin by discussing the general nature of the Russian regime and the threat it poses or could pose to nations abroad.

Since 1999, Vladimir Putin has ruled the Russian Federation, first as prime minister, then as president, then as prime minister again (nominally under President Medvedev), and then again as president from 2012. He has voiced his intention to run again for president in 2018, which, assuming he again takes office, could put him in power through 2026. During his reign,

3

most observers agree that his government has become increasingly authoritarian, corrupt, and aggressive. The main elements of national power reside in three tightly interwoven pillars—government bureaucracy, big business (either state-owned or run by Putin's oligarch allies), and the massive network of intelligence agencies. Some observers would add organized crime as an unofficial fourth pillar. The so-called *siloviki* (or 'strong-men'), consisting of Putin's closest associates, many of whom had backgrounds in the KGB and its successor, the Federal Security Service (FSB), dominate the entire structure. It is common practice for super-wealthy oligarchs to simultaneously control major corporations and serve as key ministers in the government while maintaining shadowy connections to organized crime. Dissent is discouraged—often forcefully—and challenges to the regime are thwarted through dismissal, coercion, manipulation of political processes, judicial persecution, and occasional assassination.

Sergey Markedonov, an associate professor at Russian State University, explained in a recent interview that, despite Western perspectives imagining a grand, nefarious Russian plan for expansion, most of Vladimir Putin's foreign policy moves have been reactive, not proactive. Crises arise—either from Western provocation or from local uprisings—and the Kremlin is forced to respond. Likewise, scholars from The Johns Hopkins University Applied Physics Laboratory (JHU/APL) concluded that Russia employs a "modality" rather than a strategic plan—i.e., its foreign policy has clear objectives in mind, but it pursues those vectors when crises arise and force the Kremlin to respond. Events that would trigger significant response from Russia come in three major varieties: (1) provocation from the West; (2) spontaneous local uprisings; and (3) domestic pressure from within Russia.

An objective assessment of Putin's foreign policy since 1999 would point to its reactive nature. His first war in Chechnya was sparked by the Muslim invasion of Dagestan and the string of terror attacks that followed. The Russian move against Georgia came about in the wake of Tbilisi's military move to restore its sovereignty over South Ossetia and Abkhazia. The Ukraine crisis of 2014 erupted as a local revolt against Viktor Yanukovych and his sudden volte-face in rejecting further ties to the EU. In each of these cases, Russia responded militarily, and Western critics summoned up the ghosts of Russia's tsarist and Cold War past to explain the sudden moves.

The goals of the regime include short-term objectives that are often hard to discern, and longer-term objectives that are fairly stable. Among the latter are (1) stability within the Russian Federation; (2) protection of the Putin regime; (3) strengthening Russia's prestige and respect abroad; and, tangential to this, (4) thwarting continued expansion of NATO and the European Union (EU).

Since the dissolution of the Warsaw Pact and the collapse of the Soviet Union, both NATO and the EU have marched eastward. From the Kremlin's perspective, this relentless drive to Russia's strategic periphery is vindictive, provocative, and intolerable. Russia has been forced to acquiesce as NATO reached out to embrace a unified Germany, Poland, Bulgaria, Romania, the Czech Republic, Slovakia, and the Baltic States. However, as its economy emerged from chaos and began to gain strength in Putin's early years, the Kremlin tried to deter further expansion of both NATO and the EU. Rather than backing down and recognizing Russia's great power interests in its periphery, NATO leaders continue to flirt with Georgia, Moldova, and Ukraine. Jens Stoltenberg, NATO secretary-general, recently celebrated the opening of a training base in Georgia, noting that it was a preliminary move toward accepting the country into the alliance. Likewise, Moldova has, since 1992, been edging toward a bid for membership—a move that will certainly spark a Russian reaction regarding the frozen conflict in Transnistria. Ideologues championing Putin embrace and inculcate a worldview that places Russia in the center of a Eurasian civilization built on Russian Orthodoxy and Russian culture. They demonize the West (and the United States in particular), believing (or claiming to believe) that the American CIA heads up a broad conspiracy designed to keep Russia weak and to advance Western culture and influence eastward. The regime tends toward reactionary conservatism and berates degenerate Western culture as self-destructive and unjust.

Pursuant to this worldview, the Kremlin seeks ways to increase its reach, decrease American control, promote disunity within the EU and NATO, and secure its rightful sphere of influence, especially in Eastern Europe and the Caucasus. In line with the long-term objective of keeping both the country and the Putin regime secure, Russian strategy is aimed at avoiding outright major war with the West, while at the same time striving to accomplish its objectives through aggression that will not spark a major military response. Because the regime desires to rule over the countries along its borders, it typically seeks to repress, deny, or disrupt non-Russian national autonomy ideals. The techniques used to advance the Kremlin's agenda are many and varied.

Non-military actions

There is a broad range of legal, explicit, and ethical, political, and economic activities that a nation-state can use to pursue foreign policy goals. Such methods fall under the category of routine peacetime competition. At the other end of the spectrum of conflict lies war. Between these two endpoints lie a wide range of legal, quasi-legal, and illegal activities; clandestine and covert endeavors; ethical and unethical actions. Likewise, relations between two powers most often transcend merely peace or war, and instead are characterized by various gradations of conflict. In short, modern strategy must embrace the "gray zone"—complex, often inscrutable, forms of conflict that exists between states and non-state groups. Beyond the legal and conventional methods that a great power like Russia can employ, there are a number of non-military actions available to governments and their agents and proxies.

Working with criminal elements abroad. Russian organized crime operates an extensive network throughout Russian society whose operations include extortion, fraud, cargo theft, prostitution, drug- and arms-trafficking, and other activities. The crime syndicates have controlling interests in both private and state-run businesses within Russia, and their reach extends throughout the former Soviet Union. Russian mafia elements and gangs likewise operate throughout Western Europe, North America, Latin America, and the Caribbean. A unique feature of Russian organized crime, however, is that there is no clear distinction between criminal enterprises and the government. Criminal organizations and their leadership often have direct ties to oligarchs and others in positions of power. In order to conduct business in Russia, companies often find that dealings with the government lead directly to exposure to extortion from Russian criminal networks. WikiLeaks documents exposed in the summer of 2010 accused

Russian intelligence of criminal activity including arms trafficking and working with organized crime in Spain. The alleged relationship features Russian agents offering support—money, intelligence, etc.—to criminal elements who in turn perform tasks for their patrons, thus keeping Russian agents clear of allegations of illegality.[2] Likewise, Interpol found that a variety of Russian criminal organizations, including Poldolskaya, Tambovskaya, Mazukinskaya, and Izamailovskaya, have moved into Mexico, operating through multiple small cells and engaging in a wide variety of criminal enterprises. Such criminal enterprises often have links to and support from the Russian government. The problem of Russian crime organizations is particularly severe in former Soviet states. The Russian criminal networks, however, also pose a potential threat to Vladimir Putin's regime, because they operate according to their own objectives, which may (and often do) conflict with Putin's domestic and foreign objectives.

Funding opposition (often extremist) political parties in Europe. As part of Russia's attempts to discredit those who criticize the Putin regime, the Kremlin seeks to forge links with political parties within the EU that oppose incumbent governments. Unlike political meddling during the Cold War, this new effort is not based in ideology. Indeed, Russia has supported both left-wing and right-wing populist parties.[3] Russian influence operations have targeted France, the Netherlands, Hungary, Austria, and the Czech Republic, among others. Right-wing extremist parties like Jobbik in Hungary, Golden Dawn in Greece, the Northern League in Italy, and the Front National in France are benefitting from loans originating in Russian banks.[4] Influence operations continue along a spectrum from illegal and clandestine to legal propaganda and within the "gray zones" between. In Great Britain, for example, Russia Today has broadcast programs in support of the election of Jeremy Corbyn to head the British Labour Party, championing Corbyn's resistance to economic ties with Ukraine and his opposition to Western military intervention in Eastern Europe. Russia Today likewise criticized the results of a Scottish referendum on Britain's Trident base at Faslane, suggesting the results were rigged.[5]

Economic coercion. In the early phases of the Ukraine conflict, the Russian Federation attempted to both blackmail and cajole the Kyiv government into cooperation using economic and financial power. In an attempt to popularize and prop up its political ally, Viktor Yanukovych (who had gained the presidency of Ukraine in 2010) by offering reduced prices for natural gas. In November 2013, Yanukovych, who had been pursuing closer ties with the European Union, bowed to Russian pressure and reversed course, deciding to abandon European integration in favor of closer ties to Moscow. This decision led to the Euromaidan crisis and the president's eventual ouster, but in the face of the initial popular uprising, Putin and Yanukovych signed the Ukrainian-Russian Action Plan treaty, which discounted Ukraine's natural gas purchases by a third and provided for Russia to buy up $15 billion in Ukrainian government bonds to alleviate the debt crisis. The ploy did not work, but it served as an example of how Putin would not hesitate to use his control of the Russian economy to influence events in the near abroad.

White, gray, and black propaganda. The Putin regime has demonstrated strong interest in and mastery of so-called "white propaganda"—i.e., the legal, overt use of various media to persuade targeted populations toward pro-Russian agendas. Gray propaganda originates from unclear sources. Black propaganda emanates from the opposite side that it claims to come from. Closely associated with such efforts is the practice of civil agitation to encourage opposition to the government in general or against a specific policy.

Espionage. Russian intelligence agencies include the Foreign Intelligence Service (SVR), the Main Intelligence Directorate (GRU), and the FSB. These successor agencies of the Soviet-era intelligence apparatus are active inside and outside of Russia and constitute a major pillar of the Russian government. They routinely deploy agents clandestinely to gather intelligence, particularly against states that threaten Russia, or states in which the Russian government has a foreign policy interest. Intelligence efforts include gathering information that give the Kremlin diplomatic leverage over a prospective target country, as well as spreading disinformation to sow civil discord.

Fifth columns. Russia has also demonstrated the capacity to recruit, develop, support, and control insurgent groups within target states that, during peacetime, agitate in favor of the Kremlin's policy objectives. In a war, these groups can evolve into military proxies in support of Russian military intervention.

Cyberwarfare. The Russian government has demonstrated increasingly sophisticated capability and willingness to launch cyberattacks on states that resist Russia. Following a dispute over the fate of the Soviet war statue in Tallinn in April 2007, Estonia suffered a devastating and sustained cyberattack that targeted government websites, banks and other financial institutions, the parliament, newspapers and broadcasters. The distributed denial-of-service (DDOS) attacks lasted for three weeks and then suddenly stopped. Although never legally attributed to the Russian government, most experts agree that the attacks were directed or at least coordinated with the Kremlin. In 2008, before and during the Russo-Georgian War, Moscow directed a wide-ranging cyberattack against Georgian government and news media computers, effectively rerouting the news servers to servers in Russia. The result included Russia's increased ability to manipulate the news reports of the war. Likewise, starting in 2014, Russia launched a cyberwarfare campaign against Ukraine, targeting government websites, communications centers, and other critical infrastructure. In 2016, the United States accused Russia of conducting cyberwarfare during the American presidential election campaigns, and of working with the criminal organization WikiLeaks to hack into the email accounts of American politicians and their staffs. Because current technology makes it difficult to trace the origins of cyberattacks, Russia and other countries continue to develop this line of attack against opposing states.

Military actions

Use of SPETSNAZ. Russian aggression in Crimea and Eastern Ukraine featured the widespread use of SPETSNAZ from a variety of organizations, both military and intelligence. Russia recruits SPETSNAZ domestically as well as from among populations on the strategic periphery. In the Ukraine conflict, SPETSNAZ showed up in nondescript but professional uniforms devoid of insignia, earning them the nickname "Little Green Men." The intent was to rapidly seize key installations and avoid armed conflict through preemption and intimidation.

Use of paramilitary organizations. The conflict in Ukraine also featured Russia's delivery of paramilitary organizations to bolster Russian strength while providing the Kremlin deniability. Paramilitaries included the Night Wolves (a motorcycle club), Chechen and Serbian militias, and Cossacks.

Infiltration of military supplies to insurgents. During the extended military operations in Eastern Ukraine, Russia infiltrated military supplies under the guise of humanitarian relief operations.

Deployment of battalion tactical groups (BTGs). In lieu of large-scale, army-size invasions that characterized Russian operations in World War Two and the Cold War, the Russian armed forces organized combined arms battalion tactical groups and sent them over the border to engage with loyal Ukrainian military units. As with other military actions, the Kremlin continued to deny the presence of the BTGs, even in the face of photographic evidence and firsthand accounts.

Full-scale invasion. This option, though not used in Ukraine, was the implied threat throughout the conflict. In the diplomatic exchange that went on between Russia's government and the West, Putin at one point bragged that if he wanted to, he could direct his forces to capture Kyiv within two weeks. Actual invasion, of course, entails severe strategic risk and would represent, in some ways, a failure on the part of Russia to achieve its goals through subtler means. Aimed at a NATO member, it would likely trigger major war. It is, therefore, the last resort that the Putin regime would use, but the threat of it continues to play a role.

In order to prosecute a hybrid strategy along its strategic periphery, Russia exercises escalation dominance. That is, it seeks to demonstrate its ability to rapidly threaten and carry out military operations to compel submission—potentially overwhelming targeted states with Russia's huge local military superiority. The effect of escalation dominance is that as Russian agents carry out non-military and military activities, the targeted state is deterred from responding effectively for fear of unleashing outright invasion from Russia or other harsh military actions. Membership within NATO theoretically solves the problem of Russian escalation dominance, but Putin's regime continues to test the alliance, hoping to prove that disunity and lack of resolve within NATO nullifies its effect.

ENDNOTES

[1] For a discussion of the hybrid warfare threat from Russia, see Alexander Lanoszka, "Russian Hybrid Warfare and Extended Deterrence in Eastern Europe," *International Affairs* 92, no. 1 (January 2016): 175-95.

2 Luke Harding, "WikiLeaks Cables Condemn Russia As 'Mafia State'," *The Guardian.* (2010). https://www.theguardian.com/world/2010/dec/01/wikileaks-cables-russia-mafia-kleptocracy.

3 Betina Renz & Hannah Smith, "Russia and Hybrid Warfare--Going Beyond the Label," Aleksanteri Papers, Aleksanteri Institute, University of Helsinki, Finland: Kikimora Publications (January 2016). www.helsinki.fi/aleksanteri/english/publications/aleksanteri_papers.html.

4 Peter Foster & Matthew Holehouse, "Russia Accused Of Clandestine Funding Of European Parties as U.S. Conducts Major Review of Vladimir Putin's Strategy," *The Telegraph* (18 July 2016). https://www.telegraph.co.uk/news/worldnews/europe/russia/12103602/America-to-investigate-Russian-meddling-in-EU.html.

5 Foster & Holehouse, "Russia Accused of Clandestine Funding of Euorpean Parties."

CHAPTER 2.
THE THREAT OF RUSSIAN MILITARY OCCUPATION – HISTORICAL CASE STUDIES

INTRODUCTION

This section explores the two main trends in the resistance against the Soviet occupation of Eastern Europe during the Cold War. On the one hand, a number of indigenous guerrilla movements emerged. The catalyst was the Nazi invasion of the Soviet Union in 1941. As the Wehrmacht poured across the Baltic States, Poland, and Romania, and then drove into Russia, partisans rose up to resist. In the course of four years, guerrilla organizations grew in strength and effectiveness, particularly in Poland. When the turning tide of war brought the prospect of the return of the Red Army and Soviet communism, some of the guerrilla factions resisted and carried on a lengthy insurgency. This section will examine the guerrilla operations in Latvia as an example of the course of the guerrilla resistance—its strategy, accomplishments, and ultimate failure.

Starting in 1953 just after the death of Joseph Stalin, another form of resistance arose in Eastern Europe: populist uprisings. This trend of resistance was quite different from the guerrilla movements on several counts. First, it was spontaneous. The uprisings in East Germany, Poland, Hungary, and Czechoslovakia were not planned but came about in response to perceived provocation. Second, they were urban phenomena. The guerrillas thrived in the forests; but the populist uprisings unfolded in the heart of the cities. Third, they were conspicuously non-military. The uprisings occasionally resulted in violence as mobs fought back against Soviet military crackdowns, but the protesters had no organized militias. Finally, the populist uprisings were remarkably effective. Their effects were at first apparent in the resulting shakeups within the respective communist regimes and in the concessions that the governments (and ultimately Moscow) permitted. Nevertheless, the ideological and sociological effects of the sudden uprisings reverberated throughout Europe and troubled the Kremlin deeply as the Soviet machine struggled to deal with an increasing political, psychological, and even spiritual resistance in the occupied countries.

These two strategic trends unfolded roughly sequentially. The guerrilla movements began during World War Two and continued through the mid-1950s, after which they were virtually dead. The populist uprisings began with the death of Stalin in 1953 and punctuated Cold War history until culminating with the rise of Solidarity and the end of the Soviet empire. To get a feel for the dynamics of these two strategies, this essay examines first the resistance in Latvia—centered on the guerrillas known as "Forest Brothers"—and then the resistance in Poland and the Solidarity movement.

Soviet and German Occupation, 1939-1991

Beginning in 1939, the Soviet Union, at first in collaboration with Adolf Hitler's Nazi regime in Germany, began its occupation of Eastern Europe. The Molotov-Ribbentrop Pact (also known as the Nazi-Soviet Pact) was ostensibly a treaty of non-aggression between Germany and the Soviet Union. Signed on 23 August 1939, the treaty had a secret protocol that divided Eastern Europe into German and Soviet spheres of influence. The treaty negotiations clearly anticipated subsequent military occupation of the subject regions.

Germany invaded Poland on 1 September 1939, generally recognized as the start of World War Two in Europe. Stalin, operating on the pretext of protecting ethnic Byelorussians and Ukrainians, invaded eastern Poland on 17 September, incorporating Polish territory into the Byelorussian and Ukrainian Soviet Socialist Republics (SSRs). To the north, Soviet occupations began in September as Stalin coerced the Baltic States (Estonia, Latvia, and Lithuania) to sign "mutual defense" treaties that permitted Soviet bases within their borders. The Soviet military provided the force necessary for Moscow's political agents to replace the Baltic governments with puppet communist regimes. The following June, Moscow then forced those regimes to request integration into the Soviet Union, thus formalizing Russia's occupation and de facto annexation. The Soviets likewise annexed part of eastern Finland as a consequence of the Winter War, and they annexed part of eastern Romania, creating the Moldavan SSR.

Figure 1: Soviet Occupation of Eastern Europe, 1939

This initial Soviet occupation lasted until June 1941 when Germany commenced Operation Barbarossa, the invasion of the Soviet Union. The Wehrmacht rapidly overran Eastern Europe, providing a brief window of hope for the "liberated" populations, some of whom anticipated regaining their independence under the Germans. The Nazi occupation quickly put an end to such aspirations, and the beleaguered citizens of the region suffered an equally harsh German occupation through the rest of World War Two. Following the decisive battles of Stalingrad and Kursk, the Soviet war machine steamrolled the German army and pressed on relentlessly toward the German heartland, recapturing control of Eastern Europe in 1944-45. At the end of war, the Western Allies acquiesced into peace with the Soviet Union and Stalin's effective annexation of the conquered territories. Soviet

15

rule continued for the next forty-six years, when, in 1991, the Soviet Union collapsed.

In the wake of the Cold War, Eastern European countries regained their independence and pressed Moscow to remove their troops. The weakened Russian Federation was forced to comply, and its armies eventually retreated behind their own borders. Beginning in 1999, under the increasingly authoritarian Vladimir Putin, the Russian Federation began to reassert its right to a sphere of influence along its strategic periphery. This led to a Russian invasion of Georgia in 2008 and Russia's annexation of Crimea in 2014, followed by their military and political support of Ukrainian rebels against the Kyiv regime.

Eastern Europe thus has a long historical experience of Soviet and Russian occupation. This study will focus on the dynamics of that occupation within two countries—Latvia and Poland. Both countries staged a vigorous resistance against Soviet occupation, but with different results. The lessons learned from the long, arduous, and often heart-breaking Soviet occupation provide invaluable insights as to how to counter such invasions in the future.

OTHER CASES OF RUSSIAN AGGRESSION DURING AND AFTER THE COLD WAR

Lithuania, 1991

When the Baltic states (Estonia, Latvia, and Lithuania) each declared independence from the Soviet Union, the Russian government under Mikhail Gorbachev responded by attempting to crack down on Lithuania. Military action commenced with Soviet forces seizing key government buildings and media infrastructure on 11 January 1991. They continued to assault and occupy government facilities while unarmed civilians mounted protests and demonstrations against the aggression.

On 13 January, Soviet forces moved to take over the Vilnius TV Tower. Tanks drove through demonstrators, killing fourteen, and Soviet forces began to use live ammunition against civilians. When an independent television broadcasting station managed to transmit desperate pleas to the world decrying the Soviet invasion, international pressure on Moscow mounted. This situation gave rise to a tactic that was to be repeated and refined in future interventions: denial. Gorbachev and his defense minister denied that Moscow had ordered any military action in Lithuania, claiming that the "bourgeois government" there had initiated the conflict by firing on ethnic Russians. (Coming to the defense of ethnic Russians living abroad would continue to be a favored ploy in Russian foreign policy.) Nevertheless, international and domestic reaction to the aggression caused the Soviets to cease large-scale military operations and instead use small-scale raids and intimidation.

The Soviets signed a treaty with Lithuania on 31 January, and subsequent elections saw massive popular support for independence. The Russians had been given their first post-Cold War lesson about wielding power abroad: large-scale conventional operations against sovereign states would invite unwanted scrutiny, international pressure, and domestic protest within Russia. To maintain their control over states on the periphery, they would have to employ power in a more clandestine, deniable fashion.[1]

Transnistria, 1990-92

Under Gorbachev's *perestroika* and *glasnost*, anti-Soviet sympathies grew in Moldova, and ethnic Slavs in Transnistria and Gagauzia, who favored ties to the Soviets, formed an ad hoc government that sought autonomy from the rest of Moldova. War broke out in 1992 as Moldovan forces tried to suppress separatist militias in Transnistria. To avoid the problems associated with direct military intervention, Moscow sent Cossack volunteer units to assist the separatists. For several months Transnistrian militias and Cossacks, supported by the Soviet 14th Guards Army, fought Moldovan forces, which had support from Romania.

In the summer of 1992, the remnants of the Russian 14th Army stationed in the region launched devastating artillery attacks on Moldovan forces, ending the military conflict. Transnistria became one of the so-called "frozen republics"—i.e., quasi-legal states left over from the Soviet Union.[2] The favorable outcome for Gorbachev resulted from the political strength of the ethnic Russians on the east bank of the Dniester River, the weakness of Moldova, and the strength of Russian forces still stationed in the region.

Serbian Krajina, 1991-95

Although the Russians were not directly involved in Serbia Krajina, Kremlin leaders watched with dismay as the self-proclaimed Serbian republic attempted to break away from Croatia during the latter's war for independence. Though supported and largely controlled by Serbian leader Slobodan Milosevic, Krajina's forces could not withstand Croatia's strength and determination, and the would-be republic was defeated in 1995. The Russians drew the conclusion that Western aggression against an unsupported breakaway region would prevail unless a great power (i.e., Russia) supported it with arms and diplomatic protection. When the Ukraine crisis created the Donetsk and Luhansk People's Republics, Putin and his lieutenants grew concerned that they would suffer the same fate as Serbian Krajina if Russia did not intervene.[3]

Chechnya, 1994-96

In September 1991, a coup ousted the communist government of Chechnya, the only one of the former federated states that had not come to terms with Russia as the Soviet Union dissolved. President Yeltsin attempted to put down the rebellion with Internal Troops, but the Russian forces were surrounded and compelled to withdraw. In 1993, Chechnya declared full independence from Russia. Russia began to provide funding, arms, training, and leadership to the opposition against the Chechen government, and in 1994, Russian forces joined the insurgents in two assaults on the Chechen capital of Grozny that failed catastrophically. During the campaign, Russia repeated its unconventional warfare tactics of supplying mercenary and volunteer forces, denying involvement, and using its own forces in support of the rebels. In December 1994, Russia launched an all-out invasion. Russian forces inflicted horrendous casualties among the civilian population, including those who had originally supported the intervention as well as ethnic Russians. After months of bloody fighting, the invaders finally took Grozny, but the cost in civilian life attracted universal condemnation, including from former Soviet leader Mikhail Gorbachev. The war grinded on as Russian forces advanced to try to take control of the entire country. Public confidence in Boris Yeltsin plummeted. On the last day of August 1996, the Russian government signed a cease-fire agreement with Chechen leaders, ending the First Chechen War. As in Lithuanian, Moscow learned again that the large-scale use of conventional force to impose its will on the periphery caused more problems than it solved.[4]

Dagestan and the Second Chechen War, 1999-2009

In 1999, radical Muslims from Chechnya invaded neighboring Dagestan with the aim of creating an Islamic state across the region. Russian forces intervened and expelled the invaders, but Chechen rebels responded by launching terror attacks in the region and also in Moscow. With Putin now at the helm in Moscow, Russia invaded Chechnya. Having learned hard lessons about the dangers of plunging headlong into Grozny, the Russians staged a methodical siege of the city and eventually took it before moving into the mountains to find and destroy the Muslim rebels. Following the successful conventional attack, the Russians began to pull their military forces out and instead worked with local pro-Russian proxies. The FSB and Ministry of Internal Affairs (MVD) were the agencies that directed proxy forces—an organizational technique that would continue in future wars. From 1999 through 2009, Moscow directed a sustained campaign that effectively destroyed the Islamic insurgency in Chechnya and reasserted Russian control of the region. The political and economic weakness of the Chechen government contributed to Russia's success in eliminating the rebellion by 2009. However, Putin and his advisors learned that employing poorly disciplined mass conscript armies resulted in wanton destruction, which in turn invited condemnation from abroad and from domestic opposition.

Georgia, 2008

In the early 1990s, Georgia had fought to regain control of the two breakaway regions of Abkhazia and South Ossetia, but Russian support for the separatists foiled the plan and left the two regions with de facto independence. Russian citizens with Russian passports made up the majority of the population in South Ossetia, and in the face of further attempts by Georgia to reassert control there, Putin decided to strengthen Russian control. Georgia's application for NATO membership and the fact that the Baku-Tbilisi-Ceyhan pipeline runs through the country underscored Moscow's intention to bring Georgia to heel. The situation heated up in early August as South Ossetian forces began shelling Georgian villages and Georgian forces responded. The Russians moved in more forces and began to evacuate civilians from the region. Georgian forces launched an attack into South Ossetia, initially seizing the key city of Tskhinvali. The Russians deployed units of the 58th Army along with paratroopers into the fight, and by 11 August, the Georgian forces had been expelled from the region. Russian forces then followed up with attacks into Georgia, seized the city of Gori, and threatened the capital of Tbilisi. Simultaneously they opened a second front against Georgia through operations in Abkhazia and adjacent districts. They also introduced the use of information warfare on a scale previously unseen. Russian operatives employed cyberwarfare and strong propaganda to neutralize Georgia's warfighting options and to vilify them in the press as aggressors, even accusing them of genocide. The Russian military brought journalists into the theater of war to strengthen the message of Russia protecting the population from Georgian aggression. Moscow carefully managed television broadcasts both at home and in the region, highlighting atrocities that the Georgians allegedly inflicted on the population of South Ossetia. Russian military forces performed notably better in the Georgian war than they had in the Chechen wars, in part due to a renewed reliance on professional soldiers instead of conscripts. However, strong Georgian air defenses were able to limit the use of Russian airpower, which complicated reconnaissance and the rapid deployment of Russian airborne forces. In general, Russian leaders viewed the relative success of the Georgian operation as an indicator of the need to continue modernization. Likewise, the brief campaign reiterated the key features of Russia's unconventional warfare along the periphery: (1) use of proxies when possible; (2) deniability to deflect international criticism and domestic political reaction; (3) use of information warfare, including propaganda and cyberwarfare; and (4) political preparation of subject populations and manipulation of economic conditions. All these factors would play roles in Russia's intervention in Ukraine in 2014.[5]

The Color Revolutions

The early 21st century witnessed the growing trend of popular nonviolent demonstrations and uprisings that demanded political change within authoritarian regimes. The phenomenon had precedents as early as the 1974 "Carnation Revolution" in Portugal, and the 1986 "Yellow Revolution" in the Philippines that toppled the regime of Ferdinand Marcos. At this time, Moscow's greatest concern involved the post-Cold War revolutions that occurred in former Soviet states or within the Soviet sphere. The 1989 "Velvet Revolution" in Czechoslovakia contributed to the downfall of the communist regime there. In 2000, the Serbian people's efforts to unseat Slobodan Milosevic culminated in the "Bulldozer Revolution." Milosevic was forced to resign in October, was arrested the following year, and was transferred to The Hague for prosecution. Edouard Shevardnadze was likewise forced from power in 2003 as a result of the Rose Revolution in Georgia. The following year saw demonstrations in Ukraine against the fraudulent election of Viktor Yanukovych. The resulting "Orange Revolution" culminated in new elections in January 2005 that brought opposition leader Viktor Yushchenko to power in place of Yanukovych. The "Tulip Revolution" in Kirgizstan (2005) was imitated in Belarus in the following year's abortive "Jeans Revolution" against the authoritarian regime of Alexander Lukashenko. Finally, the 2009 "Grape Revolution" in Moldova edged the communist government there out of power. Other color revolutions likewise occurred throughout the world and generally featured pro-democracy efforts against ruling regimes.

Russian analysts point to several common factors in the color revolutions: (1) student organizations; (2) non-governmental organizations (NGOs) exercising political influence;[6] (3) ubiquitous media coverage; (4) use of the Internet to spread revolutionary propaganda;[7] and (5) the government's eventual loss of control of (or at least loss of monopoly on) the state security apparatus. A key contributing factor in Georgia and Ukraine was the fragmentation and disunity of the political elites, which led to factionalism, infighting, and the development of new political parties.

Beyond these contributing factors, however, Russian leaders have insisted that the color revolutions were not spontaneous, legitimate uprisings, but rather were the product of deliberate manipulation and intervention from the United States. They likewise see these efforts as targeted against Russia. Thus, countering the color revolutions has become a major security concern among Putin's circle. To forestall future uprisings, Moscow has reached out diplomatically to authoritarian regimes, offering assistance in preventing populist movements. In a parallel effort, they have also garnered support within rightwing groups and parties in the EU and the U.S. by highlighting opposition to the problematic inclusion of East European populations into Western security and economic organizations, along with Putin's opposition to liberal positions on abortion, gay rights, and secularization. Putin is also able to use protection of the Russian diaspora as a pretext for more aggressive actions to counter democracy movements on the periphery.

ENDNOTES

[1] The Johns Hopkins University Applied Physics Laboratory (JHU/APL), *Little Green Men: A Primer on Modern Russian Unconventional Warfare, Ukraine 2013-2014* (Fort Bragg, North Carolina: United States Army Special Operations Command, 2015): 9-10.

[2] Frozen republics include Transnistria, Abkhazia, South Ossetia, Nagorno-Karabakh, Luhansk People's Republic, and Donetsk People's Republic.

[3] Koshkin, *What are the Kremlin's New Red Lines in the Post-Soviet Space?*, 2015).

[4] JHU/APL, *Little Green Men*, 11.

[5] Ibid.

[6] Jonathan Wheatley, *Georgia from National Awakening to Rose Revolution* (Burlington, VT: Ashgate, 2005): 146-47.

[7] Michael McFaul, "Transitions from Postcommunism," *Journal of Democracy* 16, no. 3 (2005), 12.

CHAPTER 3.
GUERRILLA MOVEMENTS AND LATVIA'S RESISTANCE AGAINST THE SOVIETS AND RUSSIAN FEDERATION

Latvia's Quest for Independence, 1917-1920

In order to understand how and why the Soviet and Nazi occupations came about, it is necessary to examine the context of Latvia's political situation after World War One. Prior to the initial Soviet occupation of 1939, Latvia had enjoyed twenty years of independence. At the end of World War One, on 18 November 1918, Latvia was recognized as an independent state. The creation of new "nation-states" in Central and Eastern Europe characterized the end of the war. This was facilitated by the collapse of Czarist Russia and the two subsequent revolutions that ultimately brought the Bolsheviks to power in 1917. The Brest-Litovsk Treaty of March 1918 provided a framework in which Germany and Russia would essentially carve up the provinces of Latvia, but the Bolsheviks had their hands full trying to survive and win the Russian Civil War. Concurrently, the weakening German Empire at first attempted to create an independent duchy in western Latvia, but the subsequent defeat of Germany left the Latvians with an opportunity to gain their independence. Latvian diplomat Zigfrids Anna Meirerovics simultaneously convinced the British government of the need for an independent country, and in October 1918, Foreign Secretary Arthur Balfour announced British recognition of the Latvian National Council as the government of Latvia. This precedent of British support would be long remembered and would energize Latvian partisans after World War Two. On 18 November, the Popular Council (a collection of nearly all parties in the country) announced Latvian independence. The weakened Russian government agreed to Latvian autonomy but retained control of the easternmost province, Latgale.[1]

The embryonic state had to determine what form of governance they would adopt, but there was infighting among the various parties on the right and left. The Social Democrats—originally a pro-Soviet, socialist party—eventually agreed to join the People's Council, a national unity organization led by Karlis Ulmanis and Mikelis Valters. The parties came together enough to create the Latvian Provisional Government in 1918, but the formation of the Latvian government took place in the context of ongoing conflict between the new republic of Germany and Bolshevik Russia. Both had formally relinquished their claims on Latvian territory, but in fact, both countries had interests there and armies in the vicinity. The Western Allies hoped to employ German armies to fight the Bolshevik threat, while the Latvian Communists and some Social Democrats hoped the Red Army would help them capture control of the country.[2] Thus the East-West dynamic that would later characterize the Cold War, already began to take shape, threatening the independence of the Baltic States.

In November 1918, the Bolsheviks annulled the Treaty of Brest-Litovsk, reasserting their claims in Eastern Europe. The following month, the Red Army reached Latvian territory, and the Latvian War for Independence began. The first phase—December 1918 through February 1919—witnessed a Soviet takeover and establishment of a Communist government. The Soviets under Lenin and allied with leftist Social Democrats in Latvia, launched a military offensive designed to recapture the country. This forced the Latvian Provisional Government (under Prime Minister Karlis Ulmanis) to strengthen ties with the occupying German regime—a move that alienated many in Latvia. The Red Latvian Riflemen, who had previously fought with Russia against Germany, led the attacks. In January 1919, the Red Army advanced into Riga and proclaimed the establishment of the Latvian Socialist Soviet Republic. The Bolshevik offensive continued westward and captured all but a small part of the Baltic coast in western Latvia. Peteris Stucka, a leftist Latvian politician, led the new regime and soon began a radical program of nationalizing the economy and executing class enemies. The prime targets were Germans, upper class Latvians, and peasants who resisted food confiscations. Starvation and executions claimed between 5000 and 10,000 Latvians in 1919. The repression sparked partisan movements throughout the country. The terror of the Stucka regime alienated most of the population, resulting in the eventual collapse and banning of the Communist Party of Latvia.[3]

The second phase of the War for Independence began with the arrival of Ruediger Von der Goltz, the new German commander in the western province of Courland. Von der Goltz thought little of the Latvian militias and distrusted the Provisional Government. He had no interest in Latvian independence but instead intended to lead a general offensive against Bolshevism. In April, the German-led Landeswehr orchestrated a coup against Ulmanis' government, replacing him with the more pliable Andrievs Niedra. The Germans and elements of the Provisional Army counterattacked from Courland and took Riga in May. Stucka and the

Soviets fled to Latgale, where they remained until a combined Latvian-Polish army defeated them at the Battle of Daugavpils. The Latvian Soviet government was abolished.[4]

On the diplomatic front, Latvian representatives at the Paris Peace Conference achieved some international sympathy for the country's independence but fell short of obtaining *de jure* recognition. Great Britain initially championed the Latvian cause, primarily as a way of creating a buffer against the Bolsheviks in Russia. However, as the Conference proceeded, the Allies began to retreat from their initial intentions. They largely believed that the Bolsheviks were a temporary problem and that the so-called White Russians would soon defeat them and reestablish the old order. That being the assumption, the Western Allies did not want to prejudice future relations with Russia by detaching the Baltic region from Moscow's control.

For a brief period in 1919, there were three governments in Latvia: the Provisional Government headed by Karlis Ulmanis (at one point located on a ship in the Baltic for security), the German-backed government of Andrievs Niedra, and the Soviet-backed Communist government of Peteris Stucka, which had fled to the eastern province of Latgale. When Latvian and Estonian armies advanced into the country from the north, they defeated the German Army at the Battle of Cesis, and Andrievs Niedra's government dissolved. Ulmanis returned to Riga in July. Pursuant to the Western Allies' desire to unite all parties against the Bolsheviks, the Latvians and Germans signed a truce, the terms of which included the withdrawal of the German Army (but not the forces of the Baltic Germans). The German forces under Von der Goltz ignored the terms and continued to search for a way to assert their control.

Von der Goltz' solution was to cooperate with Russia, placing his Baltic German forces under White Russian control. At first, this appeared to be a move designed to fight the Bolsheviks, but secretly the intent was to reestablish Russian control of Latvia, with guarantees that Baltic Germans would be granted Russian citizenship and the right to acquire land, thus preserving their domination of western Latvia. The German forces attacked Riga on 8 October but were stopped after seizing control of the left bank of the Daugava River. The loyal Latvian forces were growing both in numbers and experience, and they soon reversed the German gains, expelling them from Riga and forcing most German troops back to their homeland. Soon after, Ulmanis' government declared war on Germany with the aim of ending their occupation of Latvia. This move was popular among the ethnic Latvians and strengthened the government's control.[5]

The third and final phase of the War for Independence commenced in July 1919. With the help of the Poles, the Latvian Army turned its attention to the Soviet-backed Communists still in Latgale. In an extended battle near Daugavpils, the combined forces ejected the Red Army from Latvia, and the government of Peteris Stucka collapsed. Stucka fled to Russia. In January 1920, Latvia and Russia began to work toward a truce and eventually a permanent arrangement. On 11 August 1920, the Soviets and Latvians signed a peace treaty in which Russia abandoned all claims to Latvian territory and recognized the independence and sovereignty of the Latvian state. The young state failed, however, to win immediate membership in the League of Nations. Nevertheless, the following year, European nations recognized Latvia's independence, and the country was accepted into the organization. In 1922, the United States recognized Latvia.

Latvia's quest for and achievement of independence offer important lessons for East European countries facing potential Russian aggression. Geography, international politics, and the course of European history framed Latvia's struggle and ensured that its success would depend heavily on the behavior of others—both sponsors and opponents. The vicissitudes of the Western Entente at the end of World War One demonstrates that offers of help and succor must be considered pragmatically, not ideologically. The Western Allies, for all their proclaimed ideals, were war weary, unacquainted with Latvia's aspirations and history, and easily distracted from attention to the Baltics by their greater problem of containing Bolshevism. Hence, if Latvian independence could further Western Europe's strategic goals, then support might be forthcoming. If, however, the Latvian people's desire for freedom might aggravate relations with Russia or Germany, the Western Allies would as quickly abandon support for independence. Diplomatic engagement certainly played a role, but it was not the decisive factor in achieving international support. The strategic calculation that underlay the West's response to Soviet Communism was the key to their eventual support for Latvia.

The lesson to be learned is that a country's national aspirations must match the strategic ends of would-be sponsors. If they do not, neither ideology nor diplomatic engagement will suffice to win support. The unforeseen growth in Soviet power upset Western strategy, and that sudden shift in the balance of power was the key factor in the Soviet Union's eventual occupation of the Baltic States in 1939.

Latvia's Democratic Experiment, 1918-1934

The second aggravating factor that led to the Soviet occupation was the chaotic political situation within the newly independent Latvia. From its declaration of independence in 1918 through its first nationwide free election in May 1920, the country's Provisional Government labored to both win the War for Independence and rule the country—or at least that part it controlled on the ground. Meanwhile, the multi-party coalition known as the Popular Council struggled to write a constitution. Their efforts were hampered by their lack of a popular mandate. Occupying armies and the ongoing war made elections impossible. In addition, before the horrendous experience of Peteris Stucka's Communist regime, most Latvians liked what the Bolsheviks and their Latvian allies had to offer: land redistribution and a restored economy. Hence, the Popular Council's labors had to proceed without electoral backing from the citizenry.

Beginning in 1918, the Popular Council outlined its proposed constitution, calling for a liberal democracy open to all, irrespective of ethnicity. Jews, Poles, Germans, Russians, and others would be welcome into the political organization, providing they supported Latvian independence and sovereignty. The early political platform also called for respect for private property, which implied the necessary capitalist economy that a democracy required to thrive.

Less than a year after proclaiming independence, in the wake of Germany's defeat at Cesis, the Provisional Government began to organize for the nation's first election. The Law on Elections allowed an unmanageable number of tiny political parties to compete for the 150 seats in the Saiema. Though satisfying popular calls for extreme democracy, the result would be a legislature with many small parties holding, in some cases, just one or two seats. Nevertheless, it was a start down the road of

28

parliamentary democracy. Women were granted suffrage as well, a progressive step that had not yet been achieved in many of the Western democracies. The first election was held in April 1920. Eighty percent of eligible voters turned out.

Latvia's first elected parliament opened their session on 1 May 1920. They immediately began work on the "Satversme" (the Latvian constitution). The drafting committee looked to the constitutions of the United States, Switzerland, France, the Weimar Republic, Estonia, and Lithuania. Some, like Karlis Ulmanis (who had spent six years in America), were attracted to the broad presidential powers in America. Others—especially the suspicious Social Democrats—preferred the French model with a powerful legislature and a weak president. Still others liked Switzerland's direct democracy and use of popular referendum. In the end, the Social Democrats, who were worried about the ambitious Karlis Ulmanis, prevailed and the Satversme established a president elected by parliament to serve three years, although parliament could vote to dismiss him. His powers would be limited.[6]

From 1920 through 1934, the young state pursued a parliamentary democracy and enjoyed a high voter turnout. Sixteen major parties populated the political spectrum, but the Social Democrats remained the largest. Fourteen governments formed over the period, each featuring volatile coalitions. The Saeima—the national parliament—had four elections over the same period. Despite the strength and size of the Social Democrat party, its leadership refused to participate in most of the governments formed during the interwar years. Indeed, they deprecated the Latvian flag and national anthem, instead proclaiming loyalty to the Socialist International. Leaders of the party were more oriented on Marxist dogma than on Latvian national interest, and they feared that the pragmatism and compromise required to rule would harm their revolutionary ideals and alienate the leftists that they served.[7] Governing thus fell into the hands of the center-right Farmers' Union and smaller coalitions. Karlis Ulmanis and others led the party but their frustration grew with the increased polarization of the Social Democrats and with competition from smaller parties that ate away at the Farmers' Union seats in the Saeima. The spectrum of smaller parties included a wide variety of constituents, including urban middle class, right-wing anti-Soviets,

Communists, anti-Semitic nationalists, and ethnic-based parties (e.g., German, Jewish, Polish).

The consequences of Latvia's War for Independence actually worked in favor of the democratic experiment, at least in the beginning. Peteris Stucka's Communist Party was abolished, ridding the country of a potential Bolshevik fifth column. Other extremists (both right and left), along with the worst criminals, fled the country as the Germans and Russians departed. Those who were left, though by no means politically unified, were at least willing to work with each other and cooperated to some degree in the Saiema. It was a brief episode of political cooperation that failed to take root, and the demise of political compromise would morph into a major national security threat as the Soviet Union grew in strength and looked for an opportunity to reclaim its lost lands in the Baltics.

As happened in Germany, democracy came under assault due to the nationalism in Latvia. The country had a long history of occupation and exploitation from its neighbors, including Germany, Russia, Lithuania, and Poland. Each of these ethnic groups remained within the country's borders, along with Jews, who were increasingly viewed with suspicion, especially if they embraced Social Democracy. Karlis Ulmanis and others in the center and right began to call for "Latvia for Latvians" as a way of appealing to ethnic Latvians and denigrating foreigners. The nationalist trend worked against the democratization of Latvian society, because its champions consistently criticized the country's constitution. Among the discontented were the Aizsargi—the national guard—who no longer had a role in domestic security and instead occupied themselves by urging Ulmanis to march on Riga and rid the country of the democratic experiment.[8]

The Great Depression likewise ate away at democracy in Latvia. In the late 1920s and early 30s center-right politicians (e.g., Adolfs Klive, Arveds Bergs, and Edvarts Virza of the Farmers' Union Party) complained of the excessive democracy of the 1922 Satversme and called for reform. In the face of the deepening economic downturn, many saw a need for a more authoritarian president who could reset the economy and resist the centrifugal pull of foreign nationalities within Latvia.

The End of Democracy, 1934-39

In May 1934, Prime Minister Karlis Ulmanis and his allies staged a coup d'etat and established a dictatorship over Latvia. The Saiema was dismissed, political parties were banned, and the press was censored. The brief experience of Latvian parliamentary democracy was over. The reasons for its failure included problems within Latvian society itself. Some sixty percent of the population was rural, which did not augur well for the development of a successful political organization. The Latvian middle class was not well developed, and the standard of living for most citizens was poor. The regional ethnic differences were pronounced, and there was no history of a united people and country to draw from. Latgale, for example, had not previously been part of the Baltic provinces of Courland. Finally, the population had little experience with modern democracy or with the concept of self-reliance that often underpins successful representative government.[9]

Figure 2: Karlis Ulmanis After the Coup of 1934

Ulmanis was an unusual dictator in that he did not create a rubber-stamp legislature or attempt to legally justify his rule. Instead, he founded his government on the notion that a state of emergency required him to take command of the nation. He was first and foremost a Latvian nationalist, proclaiming his motto: "Latvia for Latvians," but he strictly forbade oppression of minorities and foreign ethnicities. Under his rule, the economy improved, along with literacy and education. Non-Latvian populations were encouraged officially to develop their own cultures, but through education and the economy, Ulmanis' regime aimed at assimilating them into Latvian culture.

The 1922 Constitution was set aside, and civil rights were curtailed. Political parties were banned (including Ulmanis' own Farmers' Union), and opposition presses were closed. Notwithstanding the loss of political freedoms, the Latvians enjoyed rapid economic growth under the new regime's state capitalism. While the rest of the world was still struggling to recover from the Great Depression, the Latvian gross national product (GNP) grew. Trade ties with the Soviet Union were reduced in favor of renewed trade with Britain, Germany, and the West.

From 1934 through 1939, the overriding national security problem for Latvia was the question of how best to guarantee the nation's independence. The advantage of hindsight allows the conclusion that nothing Latvia did would have prevented the disasters that soon engulfed the nation. In the uncertain and volatile situation during the interwar years, Ulmanis and his political allies pursued several approaches. First, along with sixty-one other world nations, Latvia signed the Kellogg-Briand Pact, which sought to outlaw the use of war as a means of resolving international conflict. The treaty had no practical effect on containing the conflicts that would soon give birth to World War Two, and it achieved nothing for Latvia. Second, Latvia tried to forge close ties with the countries it saw as most favorable to Latvian independence—Great Britain, France, and the United States. Although the Western powers were ideologically inclined toward self-determination, they were war-weary, geographically distant, and lacked the will and resources to oppose Bolshevik Russia. Third, Latvia expressed approval of the embryonic pact between France and Russia that would have reinforced the status quo in the Baltic States, but in the face of growing German aggressiveness and Soviet collusion with Germany, nothing came of the pact. There was also a half-hearted movement toward creating a "Baltic Entente" among Estonia, Latvia, and Lithuania that might have become a military security agreement, but again, the relative weakness of the subject powers vis a vis Germany and Russia doomed the initiative. Toward the end of the 1930s, Ulmanis' government attempted to adopt a strict neutrality in international conflict, but declarations of neutrality by minor powers were not going to be a serious obstacle to aggression from the major powers.[10]

Latvia's failure to maintain a democratic constitution was not in itself determinative regarding the country's loss of independence, but it was a contributing factor. As the Soviets cast about for a justification for their invasion of Latvia, Karlis Ulmanis' dictatorship provided an easy out. Any autocracy, no matter how benevolent, was bound to attract opponents to the regime. Lack of civil liberties meant that those enemies would be disgruntled and would likely welcome or at least cooperate with foreign intervention. Likewise, Ulmanis' dismissive attitude toward the carefully developed constitution of 1922 gave the Soviets and their Latvian allies the perfect pretext for overturning the regime.

First Soviet Occupation, 1939-41

Pursuant to the Soviet-German Nonaggression Pact of 23 August 1939, the Soviet Union was secretly granted a sphere of influence over the Baltic States, which included Finland, Estonia, Latvia, and, later, Lithuania. The so-called Molotov-Ribbentrop Pact was a brash violation of the Kellogg-Briand Pact and numerous bilateral and multilateral treaties. In a naked act of aggression, Hitler and Stalin decided to carve up Poland and allow the annexation of weaker neighboring states. The actual occupation of Latvia began shortly thereafter. On 5 October, the government of Latvia signed the Mutual Assistance Treaty. The immediate consequence of the treaty was the introduction of Soviet troops to be stationed in the country. One benefit of the treaty was renewed trade agreements between the two countries, which temporarily stimulated the Latvian economy. At the start of World War Two, Germany's interdiction of the Baltic Sea cut off Latvia from trade with Western Europe, so new opportunities for import and export were welcome. For the brief period that Latvia remained independent and Germany and the Soviet Union were cooperating, Latvia also increased trade with Germany.

As Hitler realized that the Soviets would shortly annex the Baltics, he negotiated the departure of the Baltic Germans from Latvia. This move facilitated Moscow's takeover by removing the possibility of an ethnic German population suffering oppression at the hands of the Communists, and it also divested Latvia of its best-educated and motivated citizens.

Ulmanis' government assumed the debt for the divested properties, partly in the hope that the debt would move Germany to guarantee Latvia's continued independence. It was a vain hope.

In October 1939, the Soviets entered Latvia with 25,000 troops—a figure larger than the Latvian Army—pursuant to the Mutual Assistance Treaty. Most of the Soviet bases were in the western part of the country near the Baltic coast. However, as Soviet strength grew and the possibility of a wider war increased, the Kremlin decided to annex the Baltic States outright. On 15 June 1940, Soviet NKVD units attacked three border posts in eastern Latvia, killing three guards, and the wife and son of a guard. They also abducted others. The next day the Soviets issued an ultimatum to Latvia, demanding they form a new government and allow Soviet occupation. On 17 June, the Latvian government ordered its troops to cooperate with the Soviets.[11]

Stalin sent Andrei Vyshinsky, Deputy Chairman of the Council of People's Commissars of the USSR to Riga to organize the political takeover of the country. Vyshinsky delivered Moscow's list of new cabinet members to Karlis Ulmanis, demanding that he comply. The Soviets portrayed the subsequent change in Latvia as a spontaneous people's revolution, but the entire process was supervised by the Kremlin's agent. The following month, Soviet Communist provocateurs arrived and began to organize demonstrations calling for the removal of Ulmanis and the restoration of the Constitution that he had suspended. Vyshinsky then oversaw mock national elections in which only candidates from the pro-Soviet Latvian Working People's Block were allowed to run. The results were 97.6% of the voters supposedly backed the new cabinet.[12]

In July, the Saiema voted to declare Latvia a Soviet Socialist Republic and petitioned the Supreme Council of the USSR for admission into the Soviet Union. Technically this move was unconstitutional, because the 1922 Constitution specified that such a move required a plebiscite. It proceeded nonetheless. On 5 August 1940, the USSR formally incorporated Latvia and installed Augusts Kirchensteins as president. The Western powers, including the United States and Great Britain, refused to acknowledge the new government, labeling the move as a Soviet annexation of a sovereign country. Notwithstanding, no Western powers were disposed to intervene militarily.[13]

Figure 3: Soviet Invasion and Occupation, June 1940

Working from the Soviet embassy in Riga, the NKVD immediately began recruiting collaborators and activating agents already in place. The puppet government included popular writer Vilis Lacis (Minister of Home Affairs), who had remained in contact with the banished Latvian Communist Party and the Soviets since 1928. The Chairman of the Secret Police, Vikentijs Latkovskis, was a long-time Soviet agent. The commander of the newly formed People's Army, General Roberts Klavins, had been working with Soviet intelligence since 1939.[14]

The new government set about obediently dissolving political and social organizations and removing any vestiges of a free press. Communists, regardless of education or capacity, were appointed to key cabinet positions and other offices designed to strengthen Moscow's grip on Latvian society and economy. Because the number of Latvian Communists was small—about 500 at the time of the invasion—Vyshinsky's team imported both formally exiled Latvians and a cadre of Russian Communists to oversee the local party. Moscow directed that mock elections be held, and all parties except the People's Labor Bloc were banned. Vyshynsky pushed the pre-election propaganda line that Latvian property and rights would remain intact, and he publicly shunned the idea that the USSR would annex Latvia. The rigged elections showed nearly 100% of votes for the handpicked list of candidates. This thin veneer of legality was intended to set up the next step in Moscow's incorporation of the country.

Arrest and deportation of key leaders, including Karlis Ulmanis, followed in July 1940. The newly elected Saeima sent a delegation to Moscow, where, on 5 August, the Soviet Union accepted the delegation's "request" for incorporation into the USSR. Back in Latvia, deportations increased, ethnic minorities were arrested on trumped-up charges, and Soviets took over all print media. Latvian schools were instructed to teach students "Stalin's constitution." The Latvian Army, which had been renamed the People's Army immediately after the invasion, was cut to about 11,000 men. Some twenty percent of the officers were arrested, deported, or shot, and the entire organization was supervised by Russian political commissars. By November, more than 1500 Latvians had been arrested, and an underground prison had been constructed in the basement of the Ministry of Home Affairs at 37/39 Freedom Street in Riga. For the next fifty years, it became the infamous "house on the corner" where Latvian enemies of the state were tortured.[15]

The occupation regime next moved to nationalize both farming and industry, confiscating all land exceeding thirty hectares. The financial, transportation, and commercial sectors were nationalized, and hundreds of businesses were seized throughout Latvia's major cities. Personal savings were likewise taken, and nearly all private property had been confiscated. As new Soviet and Communist functionaries moved into cities, they ejected families from their apartments and homes. Desperate to survive, Latvian citizens scrambled to buy up food and other necessities, and Latvian currency—heretofore marked by stability and strength—plummeted in value. Rationing was introduced, and anyone violating the restrictions was marked as an "enemy of the people" and arrested. Soon police were entering homes at random to inspect for surplus goods. The Soviets then imposed an artificial exchange rate of one ruble for one lat (formerly worth ten rubles), exacerbating the shortage of goods and continued to send products of all sorts back to Russia, where they fetched a much higher price. In March 1941, the lat was eliminated.

The process of Russification was begun on a small scale in 1940-41. The influx of Red Army soldiers, Communist bureaucrats, and other functionaries, along with their families, slowly increased the Russian minority. After World War Two, the process became more deliberate and featured the wholesale relocation of Russian civilians into Latvia. From the beginning of the occupation, history was Russified as well. The Latvian people were forcibly acquainted with the new facts of their past, which featured Russian benevolence and protection as the dominant theme.

Though interrupted by the German invasion of the Soviet Union in the summer of 1941, the Soviets had begun the process of collectivizing farms throughout rural Latvia. Some lower classes benefited as they were given confiscated lands in tiny portions, but the agricultural industry as a whole suffered greatly. The larger, more productive farms were either confiscated and divided up to others, or collectivized and forced to produce goods for Russia. To facilitate confiscation, many prominent landowners were deported.

There were three waves of deportations dating from the Soviets' arrival to the devastating fourth wave that occurred on the night of 14-15 June 1941. At first only prominent politicians and high-profile resisters were carted off, but when the final deportation before the German invasion occurred, 15,424 Latvian citizens were taken to Siberia, including nearly one hundred infant children. Many died during the journey. Their experiences have been well documented by the surviving victims. [16] Communist officials and police would arrive in the middle of the night and order the targeted family to pack a few belongings. They would then be marched to a railroad station and loaded onto cattle cars for the long trip to the Russian hinterland. There were three main destinations: Krasnoyarsk and Novosibirsk districts in Siberia, and Karaganda district in Kazakhstan. The Latvians, depending on their former place in society, were placed under guard in prison camps or set to work in "corrective labor camps." The specific victims and their families were chosen by local Latvian Communists and other collaborators. [17]

Latvian Resistance

As early as August 1940, Latvian citizens began organizing resistance movements against the Soviets and their puppet regime. Twenty years of independence had sown the seeds of nationalism and love for freedom, and the embryonic insurgencies that sprang up had in view the destruction of the Latvian Communist regime. In October, the "National Union of Cesis

Students" was founded. Other student movements followed—in the Aizupe School of Forestry, the Jelgava Secondary School No. 1, the Jelgava Polytechnic, and the First Secondary School in Daugavpils. Adult resistance movements included "Guards of the Fatherland," "The Young Latvians," the "Latvian National Legion," and others. The resistance published leaflets opposing the regime and organized partisan bands in the country's forests. When the Germans invaded, guerrillas launched limited attacks against the retreating Soviets.

German Invasion and Occupation, 1941-44

Soviet deportations had shocked the Latvian national conscience and almost overnight converted everyone but the collaborators to hatred of the Russian oppressors. When the German Army entered the country at the start of "Operation Barbarossa," most Latvians cheered their arrival, hoping they would liberate the country from the Soviets. The Communist regime resorted to massacring political prisoners or rapidly deporting them to Russia, where many were later murdered.

As the Soviets began to pull out, Latvian partisan groups arose to fight them. Some were motivated by a genuine desire to reestablish the state; others used the temporary vacuum of power to exact revenge on the Latvian Communists. Guerrilla forces attacked retreating Soviet forces in Riga, Valmiera, Smiltene, Aluksne, and numerous other locations. A total of 129 partisan groups operated in the summer of 1941. The question for many of them was whether the Germans were there to liberate the country or establish another occupation regime.

Latvian politicians who had survived the "Year of Horror" of the Soviet occupation spontaneously emerged and tried to set up a Latvian nationalist government. When they sought contact with the Germans, however, they were quickly disabused of their aspirations and told that such initiatives were outlawed and would be punished. Latvia's future status would be decided after the war, and in the meantime, the Latvians were to support the German occupation and war effort. The Latvian partisans were at first placed under the supervision of the German Army, reduced in size, and ordered to guard key facilities and transportation hubs. The German authorities set up Latvian self-defense groups in the villages and towns, but in August 1941, these were disbanded and replaced by police. The German commander Brigadier Walter Stahlecker directed that former partisans be used in registering and eventually exterminating the Jews, along with strengthening the German occupation.[18]

As the German Army moved east, Nazi civilian officials poured into the country. They created a complex administrative bureaucracy, but the main idea was that the Germans would appoint prominent Latvians to administer "Latvian self-government." However, the goal was not any sort of Latvian national autonomy. Rather, the self-government was aimed at relieving the Germans of the burden of dealing with local issues. The Latvians in charge were tasked by their German overlords and forced to support the war effort. While some Nazi officials—most notably the Minister for the Occupied Eastern Territories Alfred Rosenberg—gave promises of post-war autonomy, others, including Himmler, had in view the mass German colonization of the Baltics.

The Nazi Sicherheitsdienst (SD) and Security Police headed security and repression efforts in Latvia. SD agents sought to recruit Latvians to assist in the elimination of Jews and Communists throughout the country. Operational Group A was the mobile punishment unit that aimed at exterminating ideological and racial enemies, including gypsies, Jews, Soviet agents, and Communists. The unit entered Latvia on June 22 1941 and split into three groups: Liepaja, Daugavpils, and Riga. Working with Latvian appointees, the SD initiated a campaign of burning down synagogues and rounding up Communists and Jews to be shot. In addition, Reichskommissar Hinrich Lohse and Generalmajor der Polizei, Franz Walter Stahlecker oversaw the organization of ghettoes in Riga, Daugavpils, and Liepaja. Latvian Jews were herded into the enclosures and their property confiscated. The 14,000 people in the Daugavpils ghetto were liquidated in May 1942. The Liepaja ghetto was also marked for extermination, with a few survivors sent to Riga. A large concentration camp was constructed at Salaspils and operated until the Soviets reconquered the region. In all, some 70,000 Latvian Jews perished in the Holocaust.[19]

Germany recruited Latvians to fight along the Eastern Front and formed two divisions that fought major actions during the war. Both ended up defending with the retreating Wehrmacht in Latvia. One division surrendered to the American Army and the other to the Soviets. Other smaller units were likewise formed and served, mostly in Latvia. A total of over 100,000 Latvian men and boys had been organized into units and fought during the conflict, some few winning the German Iron Cross for valor.

Latvian Resistance to the Nazi Occupation

There were two distinct forms of resistance during the German occupation, 1941-44. Nationalistic Latvians who had hoped that Berlin would liberate the country and allow its reestablishment as an independent state were flatly denied their aspirations. The Nazi leaders outlawed any such movements, and they attempted to remove vestiges of Latvian patriotism. The disappointed Latvian leaders resorted to resistance that was aimed at political organization and exposing the Nazi regime's true motives.

The other form of resistance was the Soviet-backed partisan movement. Soviet aims were, of course, the reinstatement of the puppet Communist regime and the disruption of German military operations. In the historiography of the Latvian resistance movement since the end of the Cold War, Latvian nationalists insist that the Communist partisans not be considered "Latvian resistance" as other groups are called, but rather classified as "pro-Soviet collaborators." This trend reflects the factional differences that emerged as the Soviet Red Army entered Latvia in 1944, leading to the second Soviet occupation.[20]

The resistance against the Nazi occupation was affected by the pervasive knowledge that the defeat of Germany would only hasten the return of the Soviets. During the war, therefore, the Latvians refrained from large-scale guerrilla operations against the Germans and instead confined their efforts to political and cultural resistance. They protested and tried to avoid the conscription of Latvian men into the SS Legion. They likewise pushed back against Nazi economic and cultural policies and against the deportation of workers to Germany. Unlike the French Resistance and others in Western Europe, the Latvian resistance enjoyed no aid from the Western Allies. Their resources were few, and the twin enemies' regimes deployed overwhelming military strength into the country during their respective occupations. Latvian nationalists opted to distribute leaflets, newspapers, and journals protesting the German presence. Numerous groups met secretly to propagate the idea of restoring national independence after the war.

In November 1941, A. Caupals reestablished the Latvian Nationalist Union (LNU)—an organization started under the Soviet occupation but then disbanded. The LNU published flyers, facilitated communications among resisters, and collected weapons. Its clandestine network reached into the Riga police force, the city's Children's Hospital, the Red Cross Nursing School, the joint stock company "Vairogs," the Philology Faculty of Riga University, the State Technical College, and other institutions. Their illegal newspaper, *Tautas Balss* (Voice of the People) was published and distributed throughout Latvian cities. It urged the population to oppose Nazi orders and to resist both the Germans and Russians. In November 1942, Nazi agents arrested over 100 LNU members, and the newspaper was stopped.

Groups of students, intelligentsia, and other nationalists likewise pushed back against the Nazi occupiers by urging citizens toward minimal support of the German war effort. A group known as the "Patriots" insisted that they also opposed the Soviets and that the only hope for Latvian independence would be the victory of Great Britain and the United States. The old anti-Semitic extremist group known as the "Perkonkrusts" (Thunder Cross), led by Gustavs Celmins, also resisted the German regime, publishing underground newspapers calling for the restoration of independent Latvia. They called for citizens to refuse the orders of their German overlords. Celmins had, at first, collaborated with the Germans against the Soviets in the hope that they would support national independence, but when he realized that Hitler intended to colonize the Baltics, he turned against the regime. In all, there were dozens of small resistance groups—each with memberships of fifty or less—that published illegal papers and attempted to keep alive the aspiration for a restored state.[21]

What was left of Latvian political leadership in 1943 attempted to unify the disparate resistance efforts in forming the Central Council of Latvia (CCL). The organization was composed of politicians from the four most prominent parties of the last sitting Saeima: the Social Democrats, the Farmers' Union, the Democratic Centre Party, and the Latgallia Christian Farmers and Catholic Party. The underground organization met for the first time in Riga on 13 August 1943. Konstantins Cakste, a Riga University Law professor, was elected chairman. The CCL declared its purpose to be the restoration of the Republic of Latvia. They reached out to the Western Allies and reiterated the message that Latvia was not voluntarily assisting the Nazi regime. They organized committees on foreign affairs, military, information, legal, economy, resource collection, and communications. In February 1944, the CCL declared its political platform—that it was opposed to both occupying regimes, Nazi and Soviet. It called for Latvians to preserve and protect national economic and cultural assets. Notwithstanding its high ideals, the CCL was forced to flee the country in the face of the Red Army's invasion. After the war, it attempted to reassert control in Latvia, but it failed, primarily because of the postwar ramifications of the East-West standoff.

Strategic Miscalculation

Key to the CCL's perspective was the Atlantic Charter. Declared in August 1941, the Charter was the foundational Anglo-American policy statement regarding war aims. U.S. President Franklin Roosevelt sought to influence Great Britain toward a postwar international system of national freedom and global security. The British, in turn, wanted to recruit the United States into the war effort. The joint declaration that became known as the Atlantic Charter brought the two powers together with a vision for a global strategic objective. The Charter set forth that the Allied powers had no territorial ambitions but that the peoples of the world themselves would determine their own governance. Self-determination, freedom of navigation, and the lowering of trade barriers would characterize the postwar world. The two nations also called for the more idealistic goals of world disarmament and the promotion of general welfare for all.

The Atlantic Charter was at the heart of a grand strategic miscalculation by the resistance movements of Eastern Europe. The beleaguered Latvian nationalists, who had first lost their democracy in 1934 and then their independence in 1939, looked to the leadership of FDR and Winston Churchill with faith in the high-sounding words. But the Western leaders, when they spoke about national autonomy like Woodrow Wilson had done at the conclusion of World War One, had little understanding or appreciation of what that meant for Eastern Europe. When Hitler invaded the Soviet Union in 1941, the Western Allies suddenly found themselves in cahoots with their erstwhile enemy, the Communists in Russia. That relationship—vital for the purpose of defeating Nazi Germany—would change the entire direction of the postwar world, especially for Eastern Europe.

The logic of an insurgent movement deploying a guerrilla force is based on the feasibility of those guerrillas defeating or disrupting an occupying force, or at least surviving as a physical nucleus of resistance. In the case of Eastern Europe, guerrillas formed from the ranks of national military units, local militias, and bands of anti-Nazi or anti-Soviet citizens. However, from 1941 onward, those insurgents cultivated a strategic vision in which the Western Allies would triumph over Germany and, in the wake of the Nazi defeat, remake Europe based on the ideal of self-rule. The actual strategic situation was altogether different. Nazi Germany would expire on the steppes of Russia and at the hands of the Red Army. Moscow—not London, and not Washington—would dictate the fate of Eastern Europe. The prospect of the Western Allies girding themselves for a second major war against Soviet Communism in the name of providing independence for East European nations that had little cultural or historical connection to the West was illusory.

It is possible, then, to consider what might have occurred if the resistance movements of Eastern Europe had read the situation accurately, anticipated a sustained Cold War confrontation, and acted in accordance with pragmatism rather than idealism. As we will see, the guerrilla movements had all but died out by 1956, because they were achieving little and at great cost. What might have resulted if visionary insurgent leaders had instead focused on building clandestine networks aimed at non-violent resistance, political organization, labor unions, social groups, and cultural preservation? This was the direction that resistance movements eventually adopted, but only after years of failure.

The central principle that should have guided the resistance in Latvia and elsewhere was that national independence—if it were ever to be realized—would happen as a result of events outside the control of the insurgents themselves. Hindsight informs us that Latvia and the other East European countries achieved independence because the Soviet Union collapsed, not because those countries fought their way to freedom. Once the occupying power was seriously weakened, the national resistance movements did indeed exert themselves effectively and pushed the Russians out, but the main reason for the Russian retrenchment was its government's own failures and the patient persistence of the West.

Soviet Reoccupation, 1944-56

As the Soviet Red Army approached Latvia, General Janis Kurelis assumed command of a growing group of Latvian resisters who intended to fight them. The Germans permitted the group to form and intended to use it as a host nation force to oppose the Soviets, but with the Wehrmacht continuing to face defeat all along the eastern front, Kurelis and the other leaders made clear their intention to oppose both occupying powers, and they refused to cooperate with the Germans. Nazi leaders responded by cornering and destroying a large contingent of the so-called "Kurelians," deporting or executing most of its leaders. Thus, just as the Soviets were beginning to reassert their control of Latvia, one of the most effective guerrilla groups was destroyed.

Soviet-backed partisans also operated in Latvia with the aim of opposing the German war effort and paving the way for the return of the Red Army. Because most of the population had been alienated against the Soviet regime in 1940, the number of partisans was small—under a thousand. They were trained and supported by the Red Army, and starting in 1943, they began active operations against the Wehrmacht in Latvia. Their efforts were confined to sabotage and intelligence operations for the most part, and the Germans employed harsh countermeasures against them and anyone thought to be supporting them.

Pursuant to the arrangements worked out among the Allied powers at Tehran in 1943 and confirmed in Yalta in 1945, the Soviets reoccupied the Baltic States. In 1948, Stalin ordered the complete collectivization of agriculture in the subject states along with the elimination (primarily through deportation) of the landowning class and the troublesome partisan units that threatened the Soviets' monopoly on military power.[22]

The reoccupation was an expected but complicated phenomenon. Latvians who had survived the German occupation were glad to see the Wehrmacht and German officials depart, and they hoped that the return of the Soviets might lead to an end of conflict and a return to some form of normality. At the same time, the memory of 1940 was still fresh, and many took a dim view of the Russians and their Latvian allies. The Soviets, for their part, were also conflicted. They viewed the Latvian population with suspicion—especially those who had worked with the Germans. Unfortunately, they also had to remind their functionaries, including a number of criminals who victimized the population under the guise of being party officials that Latvia was and remained a Soviet Socialist Republic. Notwithstanding all these sentiments, Stalinist repression resumed with a vengeance. A recent study revealed that 2625 Latvians were arrested in 1940-41, with another 9546 (estimates as high as 15,000) deported, of whom about 2000 perished. Another 2000 were sentenced to death and killed. From 1944 through 1953, 20,000 people were sentenced to death, 88,000 were incarcerated in camps (about 10% dying there), and 41,393 (estimates range to 44,000) were deported, about 5000 of whom died.[23] Prior to the war, deportations were aimed at political elites and intelligentsia. After the war, they targeted alleged "kulaks" and "bandits" who were accused of resisting agricultural collectivization. Beyond the figures above, nearly 100,000 were placed in camps for various lengths of time.

The Latvian Resistance

Anti-Soviet partisans began to form as early as 1944. In order to avoid or escape conscription into the Red Army, and in the face of the threat of deportation, many young men fled to the forests, sometimes accompanied by their wives and children. Their intent was not necessarily to fight against the Soviets, but simply to avoid facing them. Initially, large partisan units were formed near Balvi and Vilaka in 1945, with smaller groups forming throughout other forested areas. Nevertheless, the State Security Ministry conducted effective anti-partisan sweeps of the forests, and, especially in the winter, these sweeps were able to locate and destroy the larger units. Logistical scarcity and the danger of detection therefore resulted in the multiplication of small guerrilla groups.

Various organizations attempted to organize and coordinate the guerrilla bands—some of which numbered only five to ten men. The Latvian National Partisan Association in Livland and Latgallia, the Northern Courland Partisan Organization, the Latvian National Partisan Organization in Courland, the Latvian Defenders of the Homeland Association in

45

Latgallia, and the Fatherland's Hawks in Southern Courland sought to orchestrate a sustained guerrilla resistance to the Soviets. They were successful in publishing leaflets and underground newspapers, but they fell short in prosecuting actual guerrilla operations on anything but a localized level.[24]

From 1944 through 1946, the partisan groups were most active, attacking Soviet and Latvian communist forces, police, and officials. They were occasionally effective in temporarily disrupting government control, especially in remote areas. Moreover, Soviet countermeasures combined with the growing awareness that the Western powers were not going to intervene caused a gradual decline in the numbers of partisans. Amnesty programs likewise coaxed a number of resisters out of hiding. In all, some 20,000 Latvians had gone underground in partisan bands. About 900 bands had been destroyed, comprising roughly 10,000 killed. From 1944 through 1953, they had conducted some 2659 attacks. They had killed about 1000 civilians in Latvia and killed or wounded about 1700 military and police personnel.

Soviet countermeasures included the use of locally recruited "destroyer battalions" that hunted partisan bands and attempted to bring them to battle. They also created units that masqueraded as partisans who would perpetrate crimes against civilians in an attempt to neutralize civilian support for the resisters, but the most effective countermeasure was the infiltration of informants. Partisans who were caught were often mercilessly tortured publicly for information. Occasionally, they would be executed and left in the middle of a village so that observers could watch for and capture any family members who mourned their dead.

The fate of the Latvian "Forest Brothers" and that of their other Baltic counterparts was sealed by the failure of a sustained intelligence effort orchestrated by British MI6 and the American CIA. As early as 1943, British intelligence began to establish contact with resisters in the Baltic States. They were interested not only in the continuing war effort against the Germans, but also in how the Forest Brothers might assist them in the anticipated postwar conflict with the Soviets. As the world war ended and the Cold War began, the British were chiefly interested in any plans the Soviets might have to invade Western Europe. They were determined to send agents into the Baltics to mobilize partisans there, who would in turn serve the interests of the British and later American intelligence agencies. On 15 October 1945, the Secret Intelligence Service (SIS) sent a boat from Sweden to Latvia with four agents on reconnaissance mission. Unfortunately, the boat capsized and the agents were captured and tortured. Their ciphers and radio transmitters fell into hands of Jānis Lukaševičs, a Latvian KGB officer. This began the KGB's infiltration and compromise of Western intelligence operatives in the Baltics.

In Latvia and Lithuania, KGB-directed agents continued to contact British intelligence and lured more agents to the region. The infiltrators were invariably caught, tortured, and either imprisoned or killed. In 1948, U.S. president Harry Truman tasked the newly created CIA with propaganda, economic warfare, preventive direct action, sabotage, anti-sabotage, demolition, and subversion against hostile states, including assistance to underground resistance movements. The Baltic States were an ideal spot for intelligence operations. The population was mainly anti-communist, and the partisan movements in the forest supposedly numbered tens of thousands. The region was accessible by boat and plane, and it was a forward bastion for Soviet expansion to the West. If an attack on the West were imminent, operatives there would know. The Americans' human resources were considerable, as Germany, Britain and the U.S. had many Baltic refugees who were willing anti-communists. MI6 launched "Operation Jungle," whose American counterpart was "Operation Tilestone."

The Americans set up a training camp in Germany, and the British trained other operatives in London. Their plan was to infiltrate native agents into the Baltic States to work with resistance movements and collect intelligence on the Soviets. Neither intelligence service was aware that Soviet double agents, including the infamous Kim Philby, had been operating within MI6 for years. Thus, every time the Americans or British sent agents into the region, they were caught, along with critical information about local resisters. The KGB and their subordinate agencies in Latvia and Estonia continued to send double agents to the West, where they impersonated freedom fighters ready to work with the West against the USSR. The Americans and British continued to trust the agents despite repeated reports of failure in the Baltics. Finally, when one mission sent to get radioactive waste from the Tobol River near a Soviet nuclear facility, the agents returned with a sample that, upon analysis, was found to have come from inside a nuclear reactor. In short, the KGB had planted it, and the Western spy agencies finally concluded that their operations had been compromised.

Upon further investigation, both the CIA and MI6 shut down their operations in the Baltics. They informed the disillusioned resistance in Latvia that they could no longer support them and wished them well. By 1956, all further contact was suspended. The guerrillas in Latvia, Estonia, and Lithuania had made a capital mistake in working with the Western intelligence agencies, which served only in exposing their organizations to the scrutiny of the KGB. Further, while the resistance fighters aimed for eventual national liberation, the Western intelligence agencies entertained no such ideas. They wanted to use the partisans for intelligence, knowing full well that their respective governments would not risk war merely to bring freedom to minor states in Eastern Europe.

Guerrilla movements elsewhere in the region likewise failed to achieve their goals. In Poland, the Armia Krajowa (AK), established in 1942 to fight the German occupation, grew to 150,000 (estimates up to 350,000) by 1945. When the Red Army returned to Poland in 1944-45, the AK attempted to cooperate with them while aiming for national independence at the conclusion of the war. Unfortunately, the 1944 Warsaw Uprising resulted in the near total destruction of the AK while the Red Army waited nearby, refusing to come to their aid. Other militia groups—some pro-Soviet but most anti—sprang up as the war concluded. The Red Army and the Polish communist government continued to hunt down militias and guerrilla bands through 1963, when the last ones were eliminated. Because they had no external sponsorship, the guerrilla groups had no chance of fulfilling their strategic goals. Indeed, their armed opposition served only to strengthen the grip that Moscow had on the region.

Effective resistance was thus fated to emanate from a completely different strategy—one that capitalized on social networking and non-military action.

ENDNOTES

[1] Valters Nollendorfs, Ojārs Celle, Gundega Michele, Uldis Neiburgs, & Dagnija Stasš, *The Three Occupations of Latvia, 1940-1991: Soviet and NAZI Take-Overs and Their Consequences* (Riga: Latvia Occupation Museum; 2013), 9.

[2] Daina Bleiere, Ilgvars Butulis, Inesis Feldmanis, Aivars Stranga, & Antonijis Zunda, *History of Latvia: 100 Years* (Riga, Latvia: Society Domas Spēks, 2014), 122-23.

[3] Ibid.,158.

[4] Ibid., 129-30.

[5] Ibid.,133-35.

[6] Ibid.,147-50.

[7] Ibid., 157.

[8] Ibid., 164.

[9] Ibid., 155-56.

[10] Ibid., 184-86.

[11] Nollendorfs, et al., *The Three Occupations of Latvia, 1940-1991*, 1-11.

[12] Ibid., 14-15.

[13] Ibid., 16.

[14] Bleiere, et al., *History of Latvia: 100 Years*. 242.

[15] Ibid., 245-47.

[16] See for example, Sandra Kalniete, *With Dance Shoes in Siberian Snow* (Riga: Latvijas Okupacijas muzeja biedriba, 2006).

[17] Bleiere, et al., *History of Latvia: 100 Years*, 257-59.

[18] Ibid., 262-63.

[19] Ibid., 279.

[20] Ibid., 292-93.

[21] Ibid., 295-96.

[22] Ibid., 322-23.

[23] Based on data from Russian demographer Vadim Erlikhman in a 2004 report. Bleiere, et al., *History of Latvia: 100 Years*, 347.

[24] Ibid., 362.

CHAPTER 4.
POPULIST MOVEMENTS AND POLAND'S RESISTANCE AGAINST THE SOVIETS AND COMMUNIST REGIME

INTRODUCTION

The other major form of resistance to Soviet rule in Eastern Europe was populist movements. In 1953, Joseph Stalin died, and the populations of Eastern Europe anticipated changes in the harsh Soviet rule they had endured since World War Two. Change did come, but in cities throughout the region, workers, students, intellectuals, and common people were not satisfied. A series of uncoordinated populist uprisings occurred that eventually shook the Soviet system apart.

The first significant disturbance occurred in the heart of one of the most repressive of the communist states: East Germany. In June 1953, the so-called "Workers' Uprising" broke out in Berlin. In response to the German authorities increasing work quotas by ten percent, construction workers in East Berlin spontaneously laid down their tools and refused to work. The protests widened into other sections, with the people demanding free elections and democracy. By 17 June, the demonstrations had turned into a revolt and begun to spread to other cities. Nearly half a million workers had gone on strike, and protesters began to occupy government buildings. The next day, in response to the communist government's appeals for help, the Soviet Army moved in and crushed the uprising. Several hundred people were killed, with thousands arrested. The German communist regime was purged of the members who had failed to stop the uprising earlier, and the revolt was over. The ideological implications of the Workers' Uprising were significant because the entire logic of Bolshevism was that the Communist Party was the legitimate representative of the working class. Here for the first time workers were protesting against the party and government that ruled in their name.

In 1956, the new Secretary of the Soviet Communist Party, Nikita Khrushchev, initiated a "thaw" in the Soviet system. The "secret speech" in which he revealed his intentions was leaked and became common knowledge, contributing to the popular expectation that change was forthcoming in Europe. Polish workers in Poznan gathered for mass protests in June, but again the popular uprising was put down harshly. In Hungary, 1956 saw a more sustained populist effort. The newly installed communist regime was attempting to de-Stalinize the country, but the people's desires for liberation and better economic conditions quickly outran whatever progress they could make. Tapping into Magyar nationalism and a rising sense of modernism, students and intellectuals began to march in protest in Budapest. Workers and Hungarians from all sectors of society soon joined the students, and the uprising grew to nearly 300,000 people. Police and army soldiers even joined the revolt. The people were temporarily successful as a new government was formed, the secret police dissolved, and Soviet troops withdrawn in an attempt to calm the storm. Instead, the revolt grew, and the people began to call for withdrawal from the Warsaw Pact. On 4 November, the Soviet army returned and unleashed a vicious crackdown, with thousands killed and wounded. The new government was replaced and the former prime minister executed. The new government slowly gave the people greater access to consumer goods and relaxed the harshness of former practices.

The Prague Spring of 1968 in Czechoslovakia was the next major popular uprising. The Slovak reform politician Alexander Dubcek became the head of the Czechoslovakian Communist Party in January. He intended to relax the totalitarian control imposed by former regimes in a new system he dubbed "Socialism with a Human Face," but his plans met with suspicion in Moscow. When it became clear that the Czechs intended to completely replace the communist government with a more liberal, democratic form, General Secretary Leonid Brezhnev in Moscow put his foot down. In August, Soviet and Warsaw Pact troops poured into the country to put an end to the change. The Kremlin feared a repetition of Hungary and the earlier uprisings. The Soviet move preempted any mass movement from developing, but the crackdown left feelings of betrayal and unrest in its wake. Brezhnev soon announced the doctrine named after him, the Brezhnev Doctrine, which mandated that any country which was in the Soviet communist system, must remain there.

Throughout the 1970s popular protests, demonstrations, illegal underground newspapers, and other expressions of discontent multiplied. Intellectuals continued to give voice to the growing dissatisfaction with Soviet communism. Religious and nationalist impulses likewise gained strength, even in the face of repression. Defections and escapes from Eastern Europe grew in number as well. Even the "hippy movement" made its way into the region. Police found themselves having to shut down rock concerts and arrest flower-carrying youngsters wearing jeans because the government feared this new Western influence.

Polish Populist Uprisings—the KOR and Solidarity

Poland's history is one of endured occupation. From 1772 to 1918, the nation lay divided, partitioned between Prussia, Austria, and Russia. Poles fought for their independence, even employing guerilla warfare tactics, but never met with success. It was not until after World War I that adoption of Point 13 of President Wilson's Fourteen Points secured Poland's self-determination.

During partition, and later during the Nazi occupation, Polish resistance included both violent and nonviolent means. Nonviolent resistance derived from an early 19th Century philosophical movement called "positivism," an ideological perspective that taught intellectual, cultural, and economic might was more viable than military strength.[1] These, then would become the nonviolent lines of operation that Poles would employ to resist first Nazi and then Soviet occupation.

This history is well known by Poles; it is inculcated into their national identity, and as such, made it very easy for them to connect with Solidarnosc to resist against communist oppression. For example, following the 18th century partition, Austria stifled Polish studies of history and heritage. Schoolhouse maps were devoid of depictions of Poland before partition. An underground movement known as the Agriculture Circle Society built reading rooms to continue education, established Christian stores to preserve culture, and formed credit unions to promote economic life.[2] Similarly, the People's School Society emerged in 1891 to provide education to the masses. By 1913, the society operated in 300 locations with over 42,000 members. In the German partition of Poland, not only did Poles maintain their heritage and culture through schools; students went on strike when forced to study in German or commemorate German nationalistic events.[3] In Russian-controlled Poland, a "flying university" operated in secret, graduating over 5,000 Poles in the 1880s.[4]

Following the devastation of World War Two and the imposition of Soviet communism, the Poles responded with both armed resistance (mostly in the countryside), and nonviolent resistance. The concept of nonviolent resistance through positivism was most eloquently summarized by author Maciej Barkowski, "The conspiratorial experience of organizing and running secret education became ingrained in the collective memory of national resistance."[5] Thus, the concepts that later enabled Solidarnosc were a part of Polish cultural inheritance.

While Poles had a cultural predisposition toward nonviolent resistance, the decision to resist oppression took time to mature. Before people were individually or collectively willing to overthrow their government the burden of oppression had to be endured, and they had to feel disenfranchised. After World War Two, the country became a communist state with a centrally managed economy. At first, communism seemed an effective vehicle after occupation and destruction of its infrastructure to restore the Polish people; over 6 million had died in the conflict.[6] The new government formed a construct: obedience in exchange for economic stability and security. Initially, life in Poland improved. The economy evolved from agrarian-based to more industrial-based, and peasants moved from country life to urban environs and worked on factory floors. However, this eventually led to a life of fixed prices, fixed wages, and no incentive for productivity or efficiency. This may not have been problematic, but Poles soon found themselves waiting in line for staples like bread and meat. Married couples waited years for a government-assigned apartment.

Poland, like all of the Soviet satellites, endured a managed economy. Absent the normal free market forces, this construct would eventually collapse. Periodically, the government tried to build in corrections by adjusting prices. To make these revisions more palatable a veneer of improving or changing products may accompany these increases. Sometimes the product would receive new packaging. On other occasions, it would be renamed to suggest it was a new or different item. For example, neckties were dubbed a "male hang" with an accompanying price increase.[7]

In December 1970, price increases arrived at the worst possible time, right before the Christmas holiday season. The adjustments created a detrimental effect on the overall standard of living.[8] Workers on the Baltic coast responded with strikes and demonstrations that were put down with violence. Several Poles were killed. In one infamous incident that would be revisited and memorialized, several protestors were gunned down right outside the gate of the shipyard in Gdansk.

54

In this instance, Poles were eventually mollified somewhat because First Secretary Edward Gierek replaced Wladyslaw Gomulka. This suggested the government heard their pleas and reforms would follow. Gierek's plans included garnering foreign investment and credits from capitalist countries. Then, the party encouraged personal consumptions, ironically mirroring a capitalist model. To succeed, investment and spending needed to be met with the growth, production that eventually outpaced the borrowing. Instead, due to inefficiency and recession, the influx of capital became debt.[9] For example, "licenses for machines were bought, but there were no factories to install the machines. These machines do not work to this day."[10]

Poland enjoyed an initial appearance of rising prosperity but without real reform, the economy again began to sink. For the activists, lessons learned in 1970 guided everything that followed. Jan Jozef Lipski, a founding member of KOR and later of Solidarnosc said, "1970 also did nothing but teach us that if we're not organized, we won't achieve a thing, and if the workers march separately from the intelligentsia, and the rural workers march separately, then we'll never have any impact on that power we are constantly up against. This lodged in people's minds and to this day it is paying dividends."[11]

As the economy continued to underperform, significant price adjustments again emerged as the only solution. On 24 June 1976, the prime minister proposed and the legislature accepted a general increase in food prices. There was a 69% increase in the price of fish and meat, a 30% increase in poultry, 50% in butter and cheese, 100% in sugar, 30% in flour, beans, peas, and processed vegetables. This meant near devastation for most Polish families.[12]

Strikes broke out throughout the country, but demonstrations in Radom and Ursus became the most iconic. In Radom, workers at the General Walter Metal Works refused to start the day. They did not form a strike committee but organized themselves well enough to decide to march on the Provincial Committee of the Polish United Workers Party (PUWP). This tactic became common when workers revolted in Poland. Lipski suggests, "This is a typical phenomenon under such circumstances: whenever the workers' anger boils over, the party committees alone, and not the organs of state government, are treated as the centers of power."[13] Upon discovering the officials departed in anticipation of their arrival, protestors trashed the building and set it ablaze. Looting followed.

Police began to gather evidence of the demonstrations, while constructing a clever information operation. They took photos of demonstrators, forcing them to recreate or reenact scenes of looting and wreaking havoc to use as evidence against them later and as justification for their own brutality. Around noon, Motorized Detachments of the Citizen's Militia ('ZOMO' in Polish acronym) landed at the Radom airport. These forces were resourced, trained, and equipped to counter demonstrations. Heading into action at 5:00 pm, they wrested control of the city and arrested 2000 people within two hours. They were able to do so through brutal tactics. ZOMO used clubs not only on men, but also on women, children, and the elderly. They constructed gauntlets, forcing people to endure a beating as they walked between the ranks. Several people reported that they were beaten to the point of unconsciousness.[14]

Official reporting of these atrocities was rare, perhaps because people were too scared or too frighten of reprisals. Some counted themselves lucky for having survived. At least eleven people were killed as evident by death certificates, most of which listed suicide as the cause of death. This unusually high number is contrasted by the fact that during periods of social unrest suicides typically decline, perhaps because people who are otherwise disenfranchised are able to garner hope for a brighter future.[15]

Protestors in Ursus blocked rail lines, cutting sections of the line and placing a locomotive in the gap. This prevented the flow of international commerce until the line could be restored. The police response in Ursus was so similar to that in Radom that it was clear they had prepared and were trained to respond with brutality.[16] Ten to twenty thousand people lost their jobs following the 1976 protests; police arrested 2500 people; 373 received prison sentences or fines.[17]

The Worker's Defense Committee (KOR)

As throughout their history, the Polish people would continue to resist against government oppression. Forming a group to assist the victims of Ursus and Radom became the next step. Through the act of organizing, supporting and protecting the victims and their families, and proliferating

56

information on how to continue to resist, activists could ensure a foundation for fighting against totalitarianism.

A group of the intelligentsia formed the Workers' Defense Committee known by its Polish acronym, "KOR." This group employed an imaginative design. Not surprisingly, Polish political parties and unions must register with the government. This provided means for the government to regulate their actions. First, the application could simply be denied. Second, registration required all manner of identification and reporting. Who were the leaders? Where were the offices and meeting places? Where did the funding come from? If approved, registration made the group vulnerable. They might be investigated, fined, arrested for violating laws, or simply suspended. (In the 1980s, the union called Solidarnosc was outlawed, making any and all of its activities illegal.) However, a 1930 law stipulated that committees formed to provide relief or aid, what today may be called an NGO, need not register with the government. KOR then would be an aid agency, without registration, without leaders, without a formal organization.[18] This allowed the group to be fluid in its actions.

KOR first needed to identify and connect with the victims at Ursus and Radom. On 17 July 1976, journalists, KOR members, and family members sought to observe court proceedings in Ursus, but security police blocked access.[19] Two female KOR members approached a crying woman, seeking to console her.[20] She was a victim's family member. This became a KOR tactic, sending young women to find those who were clearly suffering, then approaching them to determine if they were enduring the loss of a loved one in jail or hospital. Victims expressed the need for legal representation, finances to make up for lost wages and even babysitters to care for children as their parents' endured incarceration or convalesced. KOR collected funds, obtained lawyers to represent protestors, and even recruited boy scouts as babysitters.[21]

Though there was no formal organization, KOR still required an effective scheme of roles and responsibilities. The group built consensus during meetings, leading to decisions announced by word of mouth or pamphlets. Several subgroups, social groups, churches, and the like, maintained their identity while assisting KOR in its activities. A cadre of lawyers formed to represent the workers. As the most public of the KOR members, they invested and risked the most, exposing themselves, sacrificing time, while trying to avoid disbarment.[22] Still, no requirement was placed on any KOR member to maintain identity as a member. Individuals contributed as much or as little as their inclination or ability would allow. This ensured the entire effort was voluntary, and therefore, less susceptible to internal friction or fracture.

Included in the KOR ethos was a hard fast rule to always express and publish the truth. Anything written in a KOR document or correspondence about the state, about a government representative, if later proved to be untrue, led to a published correction.[23] This practice ensured that KOR and its message maintained legitimacy and could stand up to scrutiny by the government or rival groups.

The leadership of KOR deliberately avoided forming militias or guerrillas. Armed resistance would certainly be met with more force, and perhaps lead to a Soviet invasion similar to that of Hungary in 1956. The group renounced violence, fomenting resistance by other means. Forgiveness and reconciliation, a concept understood by Poland's large Catholic population, served as another element of the KOR platform.[24] This philosophy enabled the country as a whole to heal and move forward as the conflict ended.

Despite KOR's attempts to be unobtrusive and to be an aid agency rather than a political party, the police began targeting its members. KOR members had their vehicles vandalized, others were detained by authorities, some beaten, and still others punished by a sentencing board without opportunity to plead their case.[25] KOR encouraged all victims to report and press charges for police brutality and petitioned the government to investigate. The state responded in kind by accusing KOR as a front for espionage and its members were traitors. The police threatened accusers until they withdrew their complaints.[26]

Over time, the organization became aware of other citizens beyond Ursus and Radom that suffered under the state's oppression. To achieve their goals and remain legitimate, the organization had to support all Poles. KOR reached out to other groups seeking continued assistance, broadening the network of information and support. With the expanded mission, KOR changed its name to Committee in Defense of Human Rights and later to the Social Self-Defense Committee.[27] Though the "Workers'" moniker was dropped, the group was still commonly known as and referred to as "KOR."

Publishing became a large element of KOR's mission. The ability to proliferate ideas, share decisions of both what the group would and would not do, and to direct the actions of its members served as a natural foundation of any organization. KOR remain amorphous in this endeavor also supporting writers, publishers, those who typed and retyped the pages, and the distribution stream, but without officially connecting to any one paper or pamphlet. This meant that the authors had to clearly identify when they were writing on behalf of or in support of KOR or when their words were an editorial of their own or in support of another group or movement. Through this approach, KOR supported publication and production of over 100 documents from pamphlets to books.[28]

As KOR connected with other activist groups, new groups formed in its image. The word "solidarity" was used by "student solidarity committees" that formed to protest on behalf of those who were unjustly expelled, to create a library of forbidden books, and form a union separate from the official Socialist Union of Polish Students.[29] A new version of the "Flying University" also emerged. Lectures with titles like "On the History of People's Poland" and "Contemporary Political Ideologies" connected people to their past, encouraged continued resistance, and prepared them for their future.[30] As with other KOR activities, the police suppressed the Flying University. In one example, police raided an apartment on 11 February 1978 where Adam Michnik lectured to an audience of 120. When the students remained in defiance, the police employed tear gas to break up the class.[31] Cardinal Karol Wojtyla, who attended an underground seminary in World War Two and would later become Pope John Paul II, ensured that churches in his diocese supported Flying University courses.[32]

The Solidarity Movement

"Though we are caught in the vise of a fossilized system, a product of an outdated partition of our planet, in August 1980 we overthrew an all-powerful political taboo and proclaimed the dawning of a new era. The Polish nation achieved this as a force before the eyes of the world without threats, without violence or a drop of the opponent's blood being shed; no ideology was advanced, no economic or institutional theory: we were simply seeking human dignity."

- Lech Walesa – *A Path of Hope*

In June 1989, a Polish political party that had gained legal recognition less than two months before came to power and ended communist rule. A few months later, in a similar process Hungarians removed their communist government in October 1989. East Germans began traveling to Hungary, then crossing over the newly opened border into Austria. Later, others flowed into Czechoslovakia to take a similar journey to the West. Recognizing that they could not control this exodus, East German officials opened the border with West Germany on the evening of 9 November 1989. Before sunrise the next day, countless Germans converged on the Berlin Wall and began to dismantle it, reuniting the city.

The Cold War that had seemed interminable had ended. Countless factors led to these historical events. Among the most notable was the nonviolent resistance in Poland bearing the name "Solidarnosc" (pronounced "Solidar-nosh") or "Solidarity."

Historians variously view Solidarity as an organization of labor unions, a political party, and an insurgency. A political party seeks to participate in governance—a goal outlawed in communist Poland. An insurgency seeks to overthrow a government, supplanting it with a new form of governance. Solidarnosc began as the latter. Though possessing many of the characteristics of both a trade union and a political party, they were born outside of Poland's legal framework. Solidarnosc was an illegal organization, violating laws of assembly, press, speech, and striking in violation of labor laws. They operated as an underground with a corresponding auxiliary, but no armed guerillas. Solidarnosc employed a path of nonviolent resistance because history told them an armed insurrection would be suppressed by overwhelming force. The Poles knew violence would beget violence, and Soviet tanks would be on Warsaw streets as they had been in Budapest in 1956 or even in Prague in 1968 when even peaceful reforms sparked violent reaction. Therefore, Solidarnosc's first step was to obtain legitimacy, to become a recognized, legal body with whom Poland's communist government would negotiate. To garner this recognition, the group needed numbers—a constituency so vast that it could not be overwhelmed, intimidated, or ignored. The Polish people were able to form such a group very rapidly because they had a long history of surreptitious self-organization for resistance, armed and otherwise.

Throughout the Soviet occupation of Eastern Europe, there were episodes of protest against communism, but in each case, the government response suppressed and eventually mollified its detractors. The rise of Solidarity, however, was a fundamentally different phenomenon. The KOR had created the foundation for a broad network throughout Polish society, so that as the resistance against Soviet communism took shape in the form of Solidarity, the ground was fertile for the growth of a massive popular movement. The spontaneous uprising that would begin in the Lenin Shipyard in Gdansk would thus quickly take root and grow beyond the state's ability to control it. At the same time, the rise of a new spiritual leader within the Roman Catholic Church helped propel the Polish drive for independence.

Karol Wojtyla

Karol Wojtyla ("Carol Vwo-till-ya") entered the priesthood through an underground seminary during Poland's occupation during World War Two modeled after the "flying university." Wojtyla ascended through the Catholic Church hierarchy, receiving an appointment as Archbishop of Krakow in 1964, then as Cardinal of San Cesaro in Palatio in 1967. The Papal conclave elected Wojtyla in 1978. He was the first non-Italian pope since the sixteenth century and served twenty-seven years before he died in 2005. He became a formidable world leader and is widely recognized as having contributed significantly to the end of communist rule in Eastern Europe.

One of the themes of his sermons as the new pope traveled to his homeland was the Biblical encouragement "Do not be afraid." His spiritual instruction merged conveniently with intellectual and social trends growing within the communist bloc, lending strength especially to the Roman Catholic population of Poland. In June 1979, Pope John Paul II began a nine-day pilgrimage of his native land. His first sermon, given to an unexpected crowd of three million in Victory Square, Warsaw, clearly called the Polish people to both religious and spiritual struggle.

"My pilgrimage to my motherland …is surely a special sign of the pilgrimage that we Poles are making down through the history of the Church not only along the ways of our motherland but also along those of Europe and the world."[33]

Poles surprised even themselves at how many assembled for Mass to see and hear Pope John Paul II. Inspired by his homilies, they began to discuss how to change their plight.

Lech Walesa

Pope John Paul II could inspire and give hope to his fellow citizens, but he could not be on the ground leading change. Just over a year after the pope's visit, a strike began on 14 August 1980 in the Lenin Shipyard in Gdansk led by an out of work electrician named Lech Walesa. It originated when another shipyard worker and organizer named Anna Walentynowicz was sacked. Well loved and respected, her removal initiated protests that began in Gdansk, but then cascaded to other parts of the country. Three days later, the Interfactory Strike Committee (Międzyzakladowy Komitet Strajkowy –MKS) was established. One hundred-fifty factories throughout the nation join the ranks of the protesters—a testimony to how KOR had laid the groundwork for resistance.[34] Lech Walesa led the strike committee, which formulated twenty-one demands that the government had to accede to before the nationwide protest would end. The two most important were the right to form independent trade unions and the right to strike. All other demands derived from or supported these two.[35]

On 31 August 1980, Walesa and Deputy Prime Minister Mieczyslaw Jagielski signed an agreement that became known as "The August Agreements."[36] By holding firm until the communist government agreed to all 21 demands, Lech Walesa demonstrated to the Polish people that they could foment nonviolent resistance and through it garner self-determination.

Solidarnosc

It was after this initial protest in the summer of 1980 that disparate unions and political movements merged into a single entity named after an underground trade union newspaper - Solidarnosc ("Solidar-nosh") or

"Solidarity." Within a matter of weeks, the union was 10 million strong, consisting of 80% of the Polish workforce.[37] They elected Lech Walesa as their leader. With one large organization binding all Poles together, the government could not erode their strength, acceding to the demand of Poles in one geographic region, or one industrial field of endeavor. Similarly, smaller interest groups that were part of Solidarnosc would not be ignored, for their brethren would act in their defense.[38]

Solidarnosc acted in some ways as a typical trade union, seeking better working conditions and higher wages. It also acted like a social reform entity, promoting education and the end to alcoholism. Finally, it was clearly political, fighting for the rights of the imprisoned, and exercising a free press and free speech, especially against the communist government.[39] In the spring of 1981, Solidarnosc staged a four-hour strike including 12 million people. In the fall, they announced additional strikes to protest against government repression. Seeing the conditions emerging for Soviet intervention, the Polish government imposed martial law on 13 December 1981. The army imposed a curfew, Solidarnosc lost its recognition and again became an illegal entity, and thousands were thrown in prison including Lech Walesa. Teachers, journalists, and intellectuals critical of the system were demoted or removed. Plants and factories, already in effect nationalized, were taken over by the military.

The increased oppression would not stand. Solidarnosc continued to organize and operate using its nonviolent forms of protest. The 1980s were marked with periods of government easing restrictions, marked inability to provide for the population, and nonviolent strikes and protests. Solidarnosc garnered the attention of the outside world. Poland suffered under sanctions and food prices continued to rise. Additionally, when Mikhail Gorbachev was elected in the Kremlin, he brought reform that reverberated throughout the Warsaw Pact.

Just as they agreed in 1980, the government decided to meet with Solidarnosc starting in February 1989 in what would be known as the "round table talks." Pivotally, the communists agreed to hold elections in June. Solidarnosc won the majority of the seats and the communist party was swept from power.

Conclusion

Solidarnosc sought self-determination for all Poles. This could only be achieved by supplanting the communist government. Thus, the movement met the modern definition of an insurgency. However, Solidarnosc eschewed an armed struggle. Solidarnosc sustained illegal, nonviolent protest while seeking recognition from the government they sought to oppose by demonstrating that their movement was so popular, so populous, that its united membership was a force to be reckoned. No matter how many people the government imprisoned, no matter how many people were killed by government forces, a vast, stable element remained. Solidarnoc's showed that the disenfranchised were not a minority; rather, their suffering only remained under the threat of the force of arms.

ENDNOTES

[1] Maciej Bartkowski, "Imaging Polish Nation: Nonviolent Resistance in Poland under Partitions," *Free Russia* (June 2015): 5.

[2] Ibid., 7.

[3] Ibid., 12.

[4] Ibid., 15.

[5] Ibid., 18.

[6] *The Road to Gdansk*, directed by Maxim Ford (1983, United Kingdom, Parallax Pictures), https://www.youtube. com/ watch?v=y00Fi48VI38, 17:00 – 18:00.

[7] Jan Jozef Lipski, *KOR: A History of the Workers' Defense Committee in Poland, 1976-1981* (Berkeley: University of California Press, 1985), 30.

[8] Ibid., *KOR,* 31.

[9] Ford, *The Road to Gdansk.*

[10] Ibid.

[11] Jan Jozef Lipski, interview for Web of Stories, October 1989, interview 12782, online video, www.webofstories.com/play/jj.lipski/182

[12] Lipski, *KOR,* 32.

[13] Lipski, *KOR:* 33.

[14] Ibid., 35.

[15] Ibid., 37.

[16] Lipski, interview for Web of Stories, interview 127.

[17] Lipski, *KOR,* 41.

[18] Ibid., 44.

[19] Lipski, *KOR*, 46.

[20] Ibid., 47.

[21] Ibid., 47.

[22] Ibid., 59.

[23] Ibid., 68.

[24] Lipski, *KOR*, 68.

[25] Ibid., 88.

[26] Ibid., 103.

[27] Ibid., 198.

[28] Lipski, *KOR*, 178.

[29] Ibid., 205.

[30] Ibid., 208.

[31] Ibid., 210-211.

[32] Ibid., 212.

[33] Pope John Paul II, (Homily of His Holiness John Paul II, Victory Square, Warsaw, Poland, 2 June 1979) http://w2.vatican.va/content/john-paul-ii/en/homilies/1979/documents/hf_jp-ii_hom_19790602_polonia-varsavia.html.

[34] Koslowski.

[35] Ash, 15.

[36] Kozlowski, 4.

[37] Zein Nakhoda, "Solidarnosc (Solidarity) Brings Down the Communist Government of Poland, 1988-89," *Global Nonviolent Action Database*, September 10, 2011,

http://nvdatabase.swarthmore.edu/content/solidarno-solidarity-brings-down-communist-government-poland-1988-89.

[38] Kozlowski, 4.

[39] Kozlowski, 7.

CHAPTER 5.
OTHER CASES OF RUSSIAN AGGRESSION DURING AND AFTER THE COLD WAR

LITHUANIA, 1991

When the Baltic states (Estonia, Latvia, and Lithuania) each declared independence from the Soviet Union, the Russian government under Mikhail Gorbachev responded by attempting to crack down on Lithuania. Military action commenced with Soviet forces seizing key government buildings and media infrastructure on 11 January 1991. They continued to assault and occupy government facilities while unarmed civilians mounted protests and demonstrations against the aggression.

On 13 January, Soviet forces moved to take over the Vilnius TV Tower. Tanks drove through demonstrators, killing fourteen, and Soviet forces began to use live ammunition against civilians. When an independent television broadcasting station managed to transmit desperate pleas to the world decrying the Soviet invasion, international pressure on Moscow mounted. This situation gave rise to a tactic that was to be repeated and refined in future interventions: denial. Gorbachev and his defense minister denied that Moscow had ordered any military action in Lithuania, claiming that the "bourgeois government" there had initiated the conflict by firing on ethnic Russians. (Coming to the defense of ethnic Russians living abroad would continue to be a favored ploy in Russian foreign policy.) Nevertheless, international and domestic reaction to the aggression caused the Soviets to cease large-scale military operations and instead use small-scale raids and intimidation.

The Soviets signed a treaty with Lithuania on 31 January, and subsequent elections saw massive popular support for independence. The Russians had been given their first post-Cold War lesson about wielding power abroad: large-scale conventional operations against sovereign states would invite unwanted scrutiny, international pressure, and domestic protest within Russia. To maintain their control over states on the periphery, they would have to employ power in a more clandestine, deniable fashion.[1]

TRANSNISTRIA, 1990-92

Under Gorbachev's *perestroika* and *glasnost*, anti-Soviet sympathies grew in Moldova, and ethnic Slavs in Transnistria and Gagauzia, who favored ties to the Soviets, formed an ad hoc government that sought autonomy from the rest of Moldova. War broke out in 1992 as Moldovan forces tried to suppress separatist militias in Transnistria. To avoid the problems associated with direct military intervention, Moscow sent Cossack volunteer units to assist the separatists. For several months Transnistrian militias and Cossacks, supported by the Soviet 14th Guards Army, fought Moldovan forces, which had support from Romania.

In the summer of 1992, the remnants of the Russian 14th Army stationed in the region launched devastating artillery attacks on Moldovan forces, ending the military conflict. Transnistria became one of the so-called "frozen republics"—i.e., quasi-legal states left over from the Soviet Union.[2] The favorable outcome for Gorbachev resulted from the political strength of the ethnic Russians on the east bank of the Dniester River, the weakness of Moldova, and the strength of Russian forces still stationed in the region.

SERBIAN KRAJINA, 1991-95

Although the Russians were not directly involved in Serbia Krajina, Kremlin leaders watched with dismay as the self-proclaimed Serbian republic attempted to break away from Croatia during the latter's war for independence. Though supported and largely controlled by Serbian leader Slobodan Milosevic, Krajina's forces could not withstand Croatia's strength and determination, and the would-be republic was defeated in 1995. The Russians drew the conclusion that Western aggression against an unsupported breakaway region would prevail unless a great power (i.e., Russia) supported it with arms and diplomatic protection. When the Ukraine crisis created the Donetsk and Luhansk People's Republics, Putin and his lieutenants grew concerned that they would suffer the same fate as Serbian Krajina if Russia did not intervene.[3]

CHECHNYA, 1994-96

In September 1991, a coup ousted the communist government of Chechnya, the only one of the former federated states that had not come to terms with Russia as the Soviet Union dissolved. President Yeltsin attempted to put down the rebellion with Internal Troops, but the Russian

69

forces were surrounded and compelled to withdraw. In 1993, Chechnya declared full independence from Russia. Russia began to provide funding, arms, training, and leadership to the opposition against the Chechen government, and in 1994, Russian forces joined the insurgents in two assaults on the Chechen capital of Grozny that failed catastrophically. During the campaign, Russia repeated its unconventional warfare tactics of supplying mercenary and volunteer forces, denying involvement, and using its own forces in support of the rebels. In December 1994, Russia launched an all-out invasion. Russian forces inflicted horrendous casualties among the civilian population, including those who had originally supported the intervention as well as ethnic Russians. After months of bloody fighting, the invaders finally took Grozny, but the cost in civilian life attracted universal condemnation, including from former Soviet leader Mikhail Gorbachev. The war grinded on as Russian forces advanced to try to take control of the entire country. Public confidence in Boris Yeltsin plummeted. On the last day of August 1996, the Russian government signed a cease-fire agreement with Chechen leaders, ending the First Chechen War. As in Lithuanian, Moscow learned again that the large-scale use of conventional force to impose its will on the periphery caused more problems than it solved.[4]

DAGESTAN AND THE SECOND CHECHEN WAR, 1999-2009

In 1999, radical Muslims from Chechnya invaded neighboring Dagestan with the aim of creating an Islamic state across the region. Russian forces intervened and expelled the invaders, but Chechen rebels responded by launching terror attacks in the region and in Moscow. With Putin now at the helm in Moscow, Russia invaded Chechnya. Having learned hard lessons about the dangers of plunging headlong into Grozny, the Russians staged a methodical siege of the city and eventually took it before moving into the mountains to find and destroy the Muslim rebels. Following the successful conventional attack, the Russians began to pull their military forces out and instead worked with local pro-Russian proxies. The FSB and MVD were the agencies that directed proxy forces—an organizational technique that would continue in future wars. From 1999 through 2009, Moscow directed a sustained campaign that effectively destroyed the Islamic insurgency in Chechnya and reasserted Russian control of the region. The political and economic weakness of the Chechen government contributed to Russia's success in eliminating the rebellion by 2009. However, Putin and his advisors learned that employing poorly disciplined mass conscript armies resulted in wanton destruction, which in turn invited condemnation from abroad and from domestic opposition.

GEORGIA, 2008

In the early 1990s, Georgia had fought to regain control of the two breakaway regions of Abkhazia and South Ossetia, but Russian support for the separatists foiled the plan and left the two regions with de facto independence. Russian citizens with Russian passports made up the majority of the population in South Ossetia, and in the face of further attempts by Georgia to reassert control there, Putin decided to strengthen Russian control. Georgia's application for NATO membership and the fact that the Baku-Tbilisi-Ceyhan pipeline runs through the country underscored Moscow's intention to bring Georgia to heel. The situation heated up in early August as South Ossetian forces began shelling Georgian villages and Georgian forces responded. The Russians moved in more forces and began to evacuate civilians from the region. Georgian forces launched an attack into South Ossetia, initially seizing the key city of Tskhinvali. The Russians deployed units of the 58th Army along with paratroopers into the fight, and by 11 August, the Georgian forces had been expelled from the region. Russian forces then followed up with attacks into Georgia, seized the city of Gori, and threatened the capital of Tbilisi. Simultaneously they opened a second front against Georgia through operations in Abkhazia and adjacent districts. They also introduced the use of information warfare on a scale previously unseen. Russian operatives employed cyberwarfare and strong propaganda to neutralize Georgia's warfighting options and to vilify them in the press as aggressors, even accusing them of genocide. The Russian military brought journalists into the theater of war to strengthen the message of Russia protecting the population from Georgian aggression. Moscow carefully managed television broadcasts both at home and in the region, highlighting atrocities that the Georgians allegedly inflicted on the population of South Ossetia. Russian military forces performed notably better in the Georgian war than they had in the Chechen wars, in part due to a renewed reliance on professional soldiers instead of conscripts. However, strong Georgian air defenses were able to limit the use of Russian airpower, which complicated reconnaissance and the rapid deployment of Russian airborne forces. In general, Russian leaders viewed the relative success of the Georgian operation as an indicator of the need to continue modernization. Likewise, the brief campaign reiterated the key features of Russia's unconventional warfare along the periphery: (1) use of proxies when possible; (2) deniability to deflect international criticism and domestic political reaction; (3) use of information warfare, including propaganda and cyberwarfare; and (4) political preparation of subject populations and manipulation of economic conditions. All these factors would play roles in Russia's intervention in Ukraine in 2014.[5]

THE COLOR REVOLUTIONS

The early 21st century witnessed the growing trend of popular nonviolent demonstrations and uprisings that demanded political change within authoritarian regimes. The phenomenon had precedents as early as the 1974 "Carnation Revolution" in Portugal and the 1986 "Yellow Revolution" in the Philippines that toppled the regime of Ferdinand Marcos. Nevertheless, Moscow's greatest concern involved the post-Cold War revolutions that occurred in former Soviet states or within the Soviet sphere. The 1989 "Velvet Revolution" in Czechoslovakia contributed to the downfall of the communist regime there. In 2000, the Serbian people's efforts to unseat Slobodan Milosevic culminated in the "Bulldozer Revolution." Milosevic was forced to resign in October, was arrested the following year, and was transferred to The Hague for prosecution. Edouard Shevardnadze was likewise forced from power in 2003 as a result of the Rose Revolution in Georgia. The following year saw demonstrations in Ukraine against the fraudulent election of Viktor Yanukovych. The resulting "Orange Revolution" culminated in new elections in January 2005 that brought opposition leader Viktor Yushchenko to power in place of Yanukovych. The "Tulip Revolution" in Kirgizstan (2005) was imitated in Belarus in the following year's abortive "Jeans Revolution" against the authoritarian regime of Alexander Lukashenko. Finally, the 2009 "Grape Revolution" in Moldova edged the communist government there out of power. Other color revolutions likewise occurred throughout the world and generally featured pro-democracy efforts against ruling regimes.

. Russian analysts point to several common factors in the color revolutions: (1) student organizations; (2) NGOs exercising political influence;[6] (3) ubiquitous media coverage; (4) use of the Internet to spread revolutionary propaganda;[7] and (5) the government's eventual loss of control of (or at least loss of monopoly on) the state security apparatus. A key contributing factor in Georgia and Ukraine was the fragmentation and disunity of the political elites, which led to factionalism, infighting, and the development of new political parties.

Beyond these contributing factors, however, Russian leaders have insisted that the color revolutions were not spontaneous, legitimate uprisings, but rather were the product of deliberate manipulation and intervention from the United States. They likewise see these efforts as targeted against Russia. Thus, countering the color revolutions has become a major security concern among Putin's circle. To forestall future uprisings, Moscow has reached out diplomatically to authoritarian regimes, offering assistance in preventing populist movements. In a parallel effort, they have also garnered support within rightwing groups and parties in the EU and the U.S. by highlighting opposition to the problematic inclusion of East European populations into Western security and economic organizations, along with Putin's opposition to liberal positions on abortion, gay rights, and secularization. Putin is also able to use protection of the Russian diaspora as a pretext for more aggressive actions to counter democracy movements on the periphery.

ENDNOTES

[1] JHU/APL, *Little Green Men*, 9-10.

[2] Frozen republics include Transnistria, Abkhazia, South Ossetia, Nagorno-Karabakh, Luhansk People's Republic, and Donetsk People's Republic.

[3] Koshkin, *What are the Kremlin's New Red Lines in the Post-Soviet Space?*, 2015).

[4] JHU/APL, *Little Green Men*, 11.

[5] Ibid.

[6] Jonathan Wheatley, *Georgia from National Awakening to Rose Revolution* (Burlington, VT: Ashgate, 2005): 146-47.

[7] Michael McFaul, "Transitions from Postcommunism," *Journal of Democracy* 16, no. 3 (2005), 12.

CHAPTER 6.
OPERATION "GLADIO" AND THE RISK OF STAY-BEHIND NETWORKS

The final historical lesson to be considered is the problem of Cold War "stay-behind" networks. The Western Allies' experiences with the French Resistance and other partisan movements during World War II led postwar planners to consider how to prepare for a potential Soviet invasion of Western Europe. During the war, French partisans developed spontaneously and gradually matured into an effective guerrilla force that cooperated with Allied commanders before and after the D-Day invasion. Even so, Cold War strategists wanted to lay the groundwork for a more planned and deliberate resistance in any nation that might be the victim of a Soviet invasion. Hence, they conceived the idea of "stay-behind" forces.

The concept revolved around the idea of recruiting, training, and equipping people in each participating nation that would secretly prepare for the eventuality of a hostile occupation of their country. Supplied with weapons, ammunition, communications gear, and other essentials stored away in secret caches, these stay-behind personnel would develop networks of trusted allies within their communities who would assume the role of partisans when necessitated by war with the Soviet Union. The degree to which each nation participated varied, and the planned roles for each stay-behind network likewise differed from country to country. Some resisters would focus on performing strategic reconnaissance and maintaining communications with the West. Others would attend to the building up of a political network that would passively resist the Soviet occupation. Still others would take up arms and conduct guerrilla operations—sabotaging Soviet facilities, and battling enemy forces.

Overall command of the stay-behind operation was to fall on NATO, but the American CIA and British MI6 were allegedly involved in forming and developing the networks. The entire effort was a highly guarded secret, kept from popular scrutiny and hidden even from national legislatures. The degree to which the intelligence agencies were involved remains in dispute. What is certain is that there were stay-behind units in at least fourteen countries, including several neutral powers, most notably, Sweden, Switzerland, Finland, and Austria. The actual networks included anywhere from dozens to thousands of potential partisans preparing in secret for their wartime contingencies.

What seemed like a prudent measure to prepare in case of war gradually morphed into something more problematic. As the actual prospect of a

Soviet military invasion of the West retreated within the framework of a strategic nuclear standoff between the East and West, leaders of the stay-behind units in some countries shifted their attention from preparing for wartime resistance to manipulating peacetime politics. Again, the facts remain in dispute, but serious allegations include the charge that some stay-behind organizations allied themselves with rightwing movements, including extremist parties, who were apprehensive at the growth of leftist political parties in their countries. During the early Cold War, Communists in Italy, France, and other Western European countries grew in numbers and political strength. In response, some stay-behind networks allegedly assumed a "twofold" mission: to prepare for war and to wage a coercive terror campaign aimed at destroying or disrupting leftist political parties.[1]

The most serious allegations are that the Italian stay-behind network, originally designated "Operation Gladio," gradually involved itself in rightwing terror aimed at pursuing a so-called "strategy of tension." The intent was to perpetrate terror attacks against innocent civilians and then blame those incidents on Communists or other leftist leaders and agents. Similar allegations were made in Germany, France, and Belgium. The code name "Gladio" eventually came to mean the entire European stay-behind effort, and the term remains in common usage today.

An analysis of the Gladio controversy exceeds the scope of this work. It is certainly true that stay-behind networks were developed under the auspices of NATO, and that those networks remained hidden from the public eye. Whether some of them engaged in political conflict and terror, and the degree to which the CIA and MI6 were involved is debatable, but the very possibility impacts upon strategy formulation today. Because of Gladio and its implications, European nations are sometimes wary of developing any pre-war networks intended to act as resisters in the event of a Russian invasion. In November 1990, the European Parliament condemned the existence of stay-behind networks and called for their disbandment. The fear that such elements might engage in political warfare against their own citizens has led to some national governments disallowing any such contingency planning. In any event, strategists, commanders, and government officials must take the Gladio episode into consideration when thinking about preparing for war in a peacetime environment.

The most difficult dimension of managing preplanned resistance networks is to determine what role they should or should not play in combating Russian hybrid warfare. Because the Russian threat includes a wide spectrum of legal, quasi-legal, and illegal activities, along with both overt and clandestine operations, it can be difficult to distinguish between routine peacetime, legitimate political competition within a nation and deliberate aggression from the Kremlin. If and when Russia actually uses a targeted country's political system to threaten national security, such activities must be defended against. Stay-behind networks may be considered as part of that defense, but the political risks of such use remain high.

ENDNOTES

[1] Daniele Ganser, *NATO's Secret Armies: Operation Gladio and Terrorism in Western Europe* (London and New York: Frank Cass, 2005). Ganser's book remains controversial, and his allegations are accepted by some scholars and disputed by others.

CHAPTER 7.
CONCLUSION

CONCLUSION AND FINDINGS

The aim of this report is to address ways in which countries in Europe can buttress themselves against potential Russian aggression. Tangential to this issue is how leaders of major Western powers, and especially the United States, can contribute to this effort. What follows is an enumeration of major conclusions and findings from the referenced historical case studies, but we will preface those findings by discussing Western deterrent strategy in general.

Given the reactive nature of recent Russian aggression, Western security officials should focus on possible scenarios that begin as spontaneous local disturbances and balloon into major East-West confrontation. These potential episodes include Kazakhstan (where local leaders may try to push back against the Russian-led Eurasian Economic Union (EEU); Nagorno-Karabakh (a region within Azerbaijan that desires independence or union with Armenia); Belarus (that may attempt a move toward ties with Western economies); and Chechnya (if the region births another Muslim uprising). These regions, in addition to Moldova, Georgia, and the Baltic States, may give rise to more reactionary moves from Moscow if they threaten to disrupt Russia's strategic periphery further.

Second, there is the serious problem of Russian domestic politics. Putin has evolved an increasingly authoritarian, if not totalitarian, regime. He has put off any serious challenges to his reign through intimidation, key alliances with elites, distraction, appeals to religion and nationalism, and the occasional assassination. However, if the economy continues to deteriorate, or if his foreign adventures return serious casualties and little payoff, domestic opposition to Putin will almost certainly grow. Tyrants often resort to desperate measures to hang on to power, and military aggression can be a useful tool.

The real business of deterrence in the early 21st century revolves primarily around hybrid warfare—the use of both conventional and unconventional (irregular) forces. Hybrid warfare is characterized by the integration of military and nonmilitary considerations. Thus peaceful demonstrations, propaganda, strikes, boycotts and embargoes, sanctions, diplomatic pressuring, and information warfare play a critical role. From the perspective of the pluralistic, democratic societies of the West, popular political demonstrations and the information sphere are self-generating entities beyond the control of governments. Indeed, they are enshrined in constitutions as natural rights. However, from the perspective of authoritarian regimes, such as in China, North Korea, and Russia, they are commanded, controlled, or manipulated by the government.

Russian hawks view the "color" revolutions [1] as deliberate plots spawned by the United States and her partners with nefarious motivations aimed at harming Russia. Seemingly, spontaneous popular uprisings are seen as anything but spontaneous, and instead Russian hardliners infer conspiracy behind every blog post, newspaper article, and hacking attack. Whether Putin and his inner circle sincerely believe their own rhetoric is a matter for debate, but they invariably assign blame for such episodes on the West in general and Washington in particular.

Conversely, Russia's so-called "New Generation Warfare"—the brainchild of Russian Army Chief of Staff Valery Gerosimov—includes the deliberate deployment of non-kinetic measures as part of campaign planning. As early as the 2008 Georgian war, Moscow sent its own journalists, and in Ukraine, its own bloggers to craft a multifaceted propaganda message for both regional and worldwide consumption. Russia's new methods employ political organization, agitation, criminal violence, and other nonmilitary factors as precursors to later military intervention. Thus, when thinking about 21st century deterrence, Western policymakers have to focus on targeting their efforts against these preliminary expressions of New Generation Warfare. Whereas Cold War strategists put Soviet military deployments under a microscope in order to gain early warning against emerging threats, today's analysts must watch the Internet, the town square, and the party caucus.

Potential Russian Aggression

Putin fears and hates NATO and the Western domination that it represents. He sees the West as engaged in a relentless march eastward, deliberately targeting Russian strategic periphery. Leaders on both sides of the conflict seem ill-disposed for large-scale military operations to defeat the other, but Putin has many other tools at his disposal for having his way within his sphere of influence. Diplomatic and economic pressure (e.g., threatening to cut off natural gas and oil, boycotts, etc.), fomenting political turmoil in the Baltic States, Moldova, and elsewhere, disinformation, and cyber-attacks to disrupt military and economic activities are among his favored tactics. Russia is likewise reaching out diplomatically to potential ideological allies within the EU and even the United States. Rightwing, reactionary, and conservative groups in Europe and the U.S. are attracted to Russian propaganda against the inclusion of East Europeans into Western institutions. Many are likewise friendly to Putin's opposition to gay rights and his championing of religion as a force in society.

Along the periphery, Putin has a potential fifth column among the Russian diaspora, and Kremlin strategists are well aware of their usefulness. Putin famously lamented that the collapse of the Soviet Union left a huge problem for ethnic Russians living outside the bounds of the Russian Federation, and he has positioned himself as their champion, promising to protect their interests—with force if necessary. Thus the Russian diaspora acts as both a pretext and a motivation for Russian intervention in the near abroad.

It seems unlikely that Russia would launch a conventional attack on the Baltic States—at least until a crisis occurred that might confer legitimacy to Moscow's intervention in the region. Nevertheless, Russia has enduring interests in the Baltic region that might motivate the Putin regime or a successor to turn up the pressure and intensify the hybrid warfare operations that some already believe are occurring. These interests include: (1) using the Baltic States as a buffer between Russia and the West; (2) providing a secure overland route from Russia to its naval base at Kaliningrad; (3) protecting and exploiting the ethnic Russian (and Russian-speaking) populations, especially in Estonia and Latvia; (4) pressuring the Baltic States in an attempt to cause them to "Finlandize"—i.e., turning away from Western institutions (the EU and NATO) and instead aligning with the EEU and Russia. This last motivation may include using coercion in the Baltics as a way of indirectly intimidating other states on the periphery.

Another likely target for Russian aggression is Ukraine, which is evolving into another potential frozen conflict. While Putin is likely attempting to disengage from the conflict to avoid the drain on his resources, he and his advisors are probably content with leaving the situation in eastern

Ukraine unresolved, because it serves as a potential casus belli if they decide to use force there again. They may want to reintroduce military forces and hybrid warfare into Ukraine in order to prevent the country from strengthening ties to the West, or to intimidate border states—Moldova and Belarus in particular.

In the Caucasus region, it is likely that Putin will seek opportunities to pressure Georgia if Tbilisi continues to lean westward—particularly if NATO membership appears to be on the table. Russia would view such a move in geostrategic terms as a way for NATO to surround its Black Sea bases and extend disruptive Western influence along the strategic periphery. As demonstrated in 2008, Georgia is highly vulnerable to both outright invasion and the use of proxies in South Ossetia and Abkhazia should Tbilisi provoke the Russian regime. It is less likely that Russia will initiate a conflict with Azerbaijan or Armenia, but if regional Muslims are able to organize a serious threat to the country, it might trigger a move there.

Another likely scenario is Russian military operations against Chechen rebels, as happened twice before in recent history. The current regime in Grozny is at least nominally allied with Putin, but Russian security officials worry about the country's relations with its Muslim population. Because resulting conflict would be technically an internal affair, Chechnya has little hope of attracting international sympathy or aid in the event of war there.

Central Asia is another volatile region along the periphery that might invite Russian hybrid warfare and explicit military intervention in the near future. There is a large Russian population in Kazakhstan, and the country has been pushing against Moscow's overweening interest in keeping them firmly in the EEU, the Customs Union (CU) and the Single Economic Space (SES). Indeed, there is evidence that the two countries have not invested very much in each other's' economies, belying official statements of full integration. The regime there presides over a resource-rich market economy that does substantial business with the EU, along with its other trading partners in the EEU and the Shanghai Cooperation Organization (SCO). As long as the economic arrangements produce jobs, wage stability, and sustained investment, Moscow may continue to have its way. Unfortunately, with plummeting oil prices, Western sanctions, and the resulting Russian budget crisis, the economic consequences in Kazakhstan and elsewhere in Central Asia and the Caucasus may lead to unrest. With a conspicuous lack of political pluralism and a sustained resistance to democracy, Kazakhstani pushback may take other forms, like pressure on ethnic Russians or an attempt to go its own way in foreign affairs. Either of these scenarios could serve as a pretext for Russian aggression.

The Russian Far East seems less likely to witness aggressive moves. Indeed, Russia and China have moved closer in the wake of Western sanctions against the former and diplomatic containment against the latter. Putin is desperate to make the SCO and related economic organizations work to demonstrate a viable alternative to the EU. He is unlikely to seek confrontation with Beijing while locked in perpetual diplomatic, economic, and military struggles with the West.

Fostering Effective Counter-UW

The historical case studies referenced above suggest key lessons and conclusions that inform how a nation-state can best prepare itself to deter and defend against Russian aggression. These findings are enumerated below:

1. Historically, the cause of autonomy and independence for states in Eastern Europe depended upon the objectives and behavior of more powerful states. The subject nations were unable to defend their freedom on their own.

2. In considering the formation of an insurgency against occupying forces, it is crucial for leaders of both the insurgents and sponsoring states to align their strategic ends, ways, and means. Resistance groups typically have diverse objectives that degrade and distract the insurgency. Likewise, insurgents who are pursuing an objective of national liberation and independence must ensure that sponsoring states share that objective.

3. Insurgent leaders must develop contact with sponsoring states that go beyond the sponsor's intelligence services and reaches all the way to the head of state. Intelligence operatives who are willing and able to work with insurgents may have purposes that are not congruent with insurgent objectives. Further, intelligence leaders may not represent the sponsoring state's leadership.

4. Nonviolent resistance has a better track record than violent resistance. As demonstrated in the historical record of Soviet occupation after World War Two, the armed guerrilla movements ultimately failed to bring about change. The populist uprisings, however, became the most effective force in overturning communist rule and removing Soviet control.

5. Russian intelligence operations aim at infiltrating resistance movements, and they often do so with great success.

6. Maintaining communications with the rest of the world is a vitally important effort. Countries that may be subject to Russian aggression should take steps to ensure redundant capability. This includes continued media exposure of Russian activities and occupation, internet access, strategic communications with the government in exile, communications with the national diaspora, and communications among resistance groups.

7. Russian compatriots are a center of gravity in opposing Russian aggression. At the extreme ends of the spectrum of possibilities, they can act as a fifth column for the Kremlin, or they can spearhead the resistance against Russian aggression. Hence, they must be viewed not only as a potential problem, but also as a powerful resource, especially if they can be successfully assimilated.

8. Mobilization of all elements of society and culture in support of the resistance is crucial.

9. When preparing for resistance, the country's leadership must maintain the balance between deterring Russian aggression and preparing for it. National plans for resistance can contribute to deterrence, but they must not communicate national weakness or isolation.

10. Plans to resist illegal Russian interference in political, economic, social, or cultural affairs must not be allowed to morph into extremism, xenophobia, repression of political opponents, or terror. Transparent, representative governance coupled with law and order is the best approach.

11. Countries should withhold or withdraw visas for Russian diplomats and other agents who abuse their privileges or interfere in national politics.

12. Countries that seek to deter Russian aggression should strive to avoid polarization among political parties, because Russia aims at targeting disaffected political factions for anti-government propaganda. Every political compromise, however distasteful, can contribute to national integration and security from Russian aggression.

13. The legacy of the Operation Gladio controversy during the Cold War mandates that preparations for resistance take place with due attention to oversight by responsible officials to prevent misuse of assets in peacetime political competition. The best guarantee against another Gladio episode is redundant oversight by properly vetted government officials within the executive, legislative, and judicial branches. Preplanned resistance movements should remain clandestine but legal and safeguarded against the possibility of misuse.

It is also important to remember that any preparations for resistance are subject to use *against* Western interests in the event that a hostile government assumes control of the subject country. Likewise, it is vital to acknowledge that Russian intelligence agencies have a consistent record of infiltrating resistance organizations.

ENDNOTES

---

[1] The "color" revolutions refer to popular uprisings that occurred within the former Soviet sphere and the Balkans in the early 2000s. Some scholars include uprisings in the "Arab Spring" in this category.

APPENDIX

THE REPUBLIC OF NORTHARIA

The threat of Russian aggression in Eastern Europe is growing. The volatility of the Russian economy and the capricious nature of the Putin regime increase the risk of the Kremlin engaging in various forms of both military and non-military adventuring. In this section, we examine measures a vulnerable country must take to reduce its risk and reinforce its ability to deter, resist, and recover from attacks by the Russian Federation. We have chosen to use a fictional country to discuss the issues in order to present findings that are applicable to the entire region.

Description

The fictional "Republic of Northaria" is an East European Baltic nation-state bordering Latvia to the south and Estonia to the north. It shares its eastern border with the Russian Federation and has experienced a similar past with the other Baltic States—a history of occupation and domination by both Germany and Russia, along with an extended post-World War Two rule by the Soviet Union.

Fostering Effective Defense in the Republic of Northaria

In this section, we will describe the history and current characteristics of Northaria and simultaneously examine optimal strategies for protecting the country from various forms of Russian aggression.

Early History

Similar to its bordering neighbors, Estonia and Latvia, Northaria's political history is replete with shifting borders. In the 13th century, Northaria, sandwiched between modern day Estonia, Latvia, and Russia, fell prey to Germanic crusaders, the Teutonic Knights, seeking to Christianize the pagan population of the Baltics. During the Northern Crusades, the Germans established the Northarian capital city, Mestauskal, along with the Latvian capital city Riga. Mestauskal became part of the Hanseatic League during this period. The Livonian Order, a branch of the Teutonic Order, ruled most of Northaria until their power was challenged

88

by imperial rulers, including the Polish-Lithuanian Commonwealth, the Swedish Empire, and the Russian Empire. Germanic influence, bolstered by the considerable number of Baltic Germans remaining in the region, remained a powerful force in Northaria for some time. After the Reformation, for instance, the ethnic Northarians adopted Lutheranism as their primary religion. Northaria remained under Swedish rule for several hundred years until Russia defeated Sweden in the Great Northern War. Northaria was absorbed into the large Russian empire, where it flourished for a time. Northaria, alongside Latvia, became the Empire's nascent industrial hub.

After World War I, Northaria, for the first time in millennia, successfully pressed for independence. The newly established USSR, which was consumed with consolidating its own power, recognized Northaria's independence in the Treaty of Mestauskal. The Treaty followed a fierce, if brief, War of Independence. The nationalist trend among the titular ethnic group, the Northarians, mirrored similar trends in Europe, such as the nationalist movement in Ireland that resulted in the country's independence from Great Britain. When Northaria declared its independence political elites, mostly Northarians, formed the Constituent Assembly of 1920, drafting the new state's first constitution, which was ratified via referendum the following year. Its current borders, shared with Latvia, Estonia, and Russia, were settled during this time as well. The proceeding decade was a tumultuous one, but Northaria, unlike Estonia and Latvia, escaped the threat of coup d'état, maintaining its constitution until 1940 when it was taken over by the Soviet Union.

Independence

Northaria is a parliamentary democracy established in 1990 following its declaration of independence from the Soviet Union on April 18, 1990. The ruling body at the time, the Supreme Council of the Republic of Northaria (SSRN), ratified the declaration. In early 1991, the SSRN held an independence referendum that voters approved by a margin of over 90%. The country's independence came after 50 years of Soviet rule. Like the other Baltic States, Northaria became part of the Soviet Union after Moscow occupied the region under the Molotov-Ribbentrop Pact. The Pact formed a non-aggression agreement between the Soviet Union and Germany while also dividing the border territories between them into two separate spheres of influence. Under the auspices of this agreement, the Soviet Union annexed Estonia, Latvia, Lithuania, and Northaria in 1940. After the annexation, Northaria became a Soviet Socialist Republic of the Soviet Union, ruled by the Communist Party of Northaria until 1990. The Soviet Union officially accepted the Baltic States' independence in September 1990.

Shortly after its independence, Northaria assembled a constitutional convention. The Constitutional Assembly, as the drafting body was called, included 20 appointed representatives from the SSRN by the governing bodies of Northaria at the time. After convening for nine months, which included 40 sittings, Northaria ratified its current constitution via a general referendum in 1992. The constitution, like others in the Baltic States, establishes the continuity of Northaria with its former period of sovereignty in the period 1920 – 1940. Northaria's first constitution was ratified in 1921 following the country's declaration of independence.

The country's current constitution, while it takes some elements from the 1921 constitution, is considered a separate document. The constitution established the Majvendi, Northaria's parliament, which is the unicameral legislative body of the country. During the Soviet occupation, Northaria had a unicameral legislature, although it had virtually no independent legislative authority. Unicameralism was a common feature in Soviet republics. As a result, some former Soviet republics, including Northaria, retained the feature. Additionally, unicameralist legislatures are often more efficient lawmaking bodies since the potential for legislative deadlock between two chambers is eliminated. Some individuals in the Northarian Constitutional Assembly also argued that a unicameral legislature cost less and reduced the country's bureaucratic footprint. The Majvendi has 86 seats that are elected through proportional representation in the multi-party system, four-year election cycles. However, the threshold for representation for competing parties is at least 5% of the vote. No one political party has successfully formed a majority government in Northaria since the country first held elections in 1992. Instead, multiple parties have formed coalitions that develop sufficient blocks of power to govern the country. While multi-party systems like Northaria's are praised for consensus-based governance and increasing the political legitimacy of the government, critics point out that internal power struggles, particularly during times of crisis or unrest, are likely to emerge and fracture the political regime.

The Northarian executive branch is divided into two positions: the prime minister and the president. The president is the commander-in-chief of the armed forces and head of state, but otherwise exercises less governmental authority than the prime minister. As the head of state, the president appoints diplomatic representatives and represents Northaria in interactions with other states, such as in trade negotiations. However, all international agreements are subject to legislative ratification. The prime minister is selected by the coalition government after the election cycle and approved by the Majvendi. Once the prime minister is affirmed, he or she appoints the president, informally from a list of candidates provided by party leaders in the coalition. The prime minister also appoints members of the ministerial cabinet and acts as the head of government.

Election cycles in Northaria occur every four years in the month of April. Votes are allocated in an open list proportional representative system. This means that on each ballot, people vote for the party of their choice but also have the option to vote for the party candidates of their choice. Only parties that breach the 5% vote threshold gain seats in the Majvendi. The age of suffrage is 18.

Historical Development of Political Parties

The political parties in Northaria first became active in 1989 when then Soviet leader Mikhail Gorbachev initiated perestroika reforms. Part of the reforms included contested elections for the USSR Congress of People's Deputies, a new legislative body intended to reform the moribund political system and isolate Gorbachev's conservative opponents in the Communist Party of the Soviet Union (CPSU). Another election occurred in 1990 for new Supreme Councils. Each election saw independent candidates and parties, outsides the ruling Communist elite, participating. Although the Communist elite had a resource advantage, many of the independent candidates and parties were successful. The political opportunity provided by the elections allowed a new class of leaders and politicians to emerge with the requisite skills in attracting and mobilizing constituents to successfully steer their new countries towards a viable independence. For the most part, Northarians rejected the former Communist parties that ruled the country during the Soviet era. Most of the leaders of the parties were Russian and the Northarian population associated the parties with foreign occupation. The Northarian Communist party collapsed along with the USSR.

In Northaria, as well as the other Baltic States, three main blocs formed the independence movement. Most, if not all, of the current political parties in Northaria have connections to one or more of the major blocs. The first was a Popular Front that acted as an umbrella organization covering broad range of societal actors that cut across numerous sectors of society. The Popular Front included moderate Communists, nationalists, Greens, and Social Democrats. The interests of each sector converged only on the issue of independence; otherwise, the actors had little common ground. Not surprisingly, after the Popular Front achieved independence, the fissures became acute, leading to the formation of numerous political parties with divergent viewpoints on the future of Northaria. One of the issues surrounded the question of citizenship, which will be discussed in greater detail in the Social section. Other areas of divergence included viewpoints on the economy, particularly as it related to the desirability of free market reforms. Social welfare policies also proved contentious. The parties that first emerged from the Popular Front bloc included the National Progress Group (NPG), the Liberal Green League (LGL), and the International Equilibrium Party (IEP).

The Inter movement was another of the three blocs. It contended against the Popular Front's efforts to secure Northaria's independence from the Soviet Union. The Inter movement consisted primarily of ethnic Russians. Most were drawn from active or retired Soviet military personnel, hardline Communists, recent Russian immigrants, and some multi-generational Russian immigrants. However, some ethnic Russians were not part of the Inter movement and supported the democratic and economic reforms proposed by the Popular Front. The Collective Freedom Group (CFG) and the Socialist Unionist Coalition (SUC) were among the first parties to emerge from the Inter movement bloc in post-independence Northaria.

Alongside the Popular Front and the Inter movement was the radical nationalist Congress movement. Like the Popular Front, the Congress movement wanted to establish independence from the Soviet Union. However, after its independence, it sought to reconstitute Northaria as it was during the pre-1940 era. For political actors in this movement, that meant lustrating all former communists and criminalizing their behavior. Some advocated trying Communists as war criminals for crimes committed during the Soviet occupation of Northaria. Among the first political parties established in the wake of Northarian independence, the Patriotic Fatherland Union (PFN) emerged from the Congress movement. The PFN had some successes during the first free elections held in post-independence Northaria, but quickly factionalized into numerous other parties that held similar views.

Political Parties

Political parties are arguably the most important political actors in Northaria today. During the transition to independence, parties played a critical role in consolidating democracy in the country by following the "rules of the game" established by country's constitution. The parties' scrupulousness helped to legitimize Northaria's government among its citizens. Despite the important role that parties continue to play in the country's political system, the public has invested little trust in the parties themselves. In Northaria, as in most of the Baltic States, political parties have a low level of institutionalization. This means that the parties are prone to internal struggles and factional divisions, contributing to a cycle of party splinters, disbandment, and formation. Northarians generally do not have a strong sense of identification for one party over another, so there is only a weak link between the parties and their constituents.

The parties in Northaria, while less volatile than Latvia and Lithuania, have not been able to reach the same level of stability as those in Estonia. The volatility of parties is measured according to the entry of new parties and the exit of old parties. In Northaria, the number of new parties has decreased since its height in the 1990s. The number was higher in the 1990s as more ethnic Russians became citizens, leading to the formation of more parties that accommodated their interests or those opposing increases in Russian citizenship in Northaria. Additionally, in the early 2000s, tightened regulations regarding the entry of new parties made it more difficult to establish them. About 80% of voters reserve votes for established parties, with the remaining going to new parties. Voter turnout has also declined considerably. In the first national election, 79% of eligible voters took part in the election. Now, voter turnout is around half of eligible voters. According to the 2014 Eurobarometer survey, only around 15% of Northarians find the political parties trustworthy. The low levels of trust contribute to volatility since many voters are likely to shift support from on party to another. Moreover, while Northarian parties are more volatile than those in Western Europe, the core policies among the parties tend to not dramatically shift. The bedrock policies of support for the EU and NATO are more or less unchallenged. Moreover, most new parties are generally reconfigurations, alliances, or splinters of existing parties. The same cast of characters, policies, and funding mechanisms are evident; truly new parties are rare.

Table A-1: Parties in the Majvendi and the European Parliament[1]

Party	Majvendi Seats	European Parliament Seats
National Liberal Party	32	2
Advanced Future Party	17	1
Industrial and Farmers Fatherland Party	14	1
Rodina Patriotic Party	12	0
Moderate Conservative Coalition	11	0
Lawful Communion Coalition	9	0
Modern Preservation Party	7	0

National Liberal Party – Governing Coalition

Since 2003, the National Liberal Party (NLP) has consistently held more seats in the Majvendi than any other party has. Despite its popularity, it has not yet been able to form a majority government, usually acquiring around 30% of the votes. In 2010, the NLP formed a coalition government with the Advanced Future Party (AFP), which generally gains a smaller percentage of the vote. In the last election, held in 2014, the NLP and the AFP did not receive sufficient votes to form a coalition government. As a result, the parties invited the far-right nationalist party, the Industrial and Farmers Fatherland Party (IFFP), to form a coalition government.

The policy platform of the NLP is conservative liberalism. Since its formation, the party has advocated for free market policies. Its leaders regard integration with the European Union the cornerstone of its economic policies. Since the election in 2014, the NLP has voiced strong support for the Transatlantic Trade and Investment Partnership (TTIP), the EU's proposed trade partnership with the United States. Its policies have also lowered corporate tax rates, which hover near zero for re-invested profits; in recent years, it has lobbied hard for a flat income tax around 20%. The NLP has successfully partnered its liberal economic policies with social conservatism. For several years, its policies in this regard have shifted further to the right. While a supporter of the EU, the NLP has voiced criticism of the EU's greater rights for the LGBTQ community and marriage equality. Some NLP members have even quietly voiced support for Putin's anti-homosexual agendas in Russia. It has come down harder on support for upholding Northarian culture and traditions, evident in its support for stricter immigration controls. Under the leadership of the NLP, Northaria restricted the inflow of refugees to 530 for 2017. Some observers fear the swing towards nationalism will compromise the NLP's support for the EU.

Advanced Future Party – Governing Coalition

The AFP, alongside the NLP and the IFFP, is part of the governing coalition of the Majvendi. Since its formation in 2004, the AFP's policy positions have remained center-left, supporting social welfare provisions for Northaria. Its platform includes support for anti-corruption measures and public financing for elections. While not anti-business, the AFP has spoken out against the austerity measures that have reduced budget allocations to education and health care. In past years, it has also voiced support for further strengthening trade unions. Socially, the AFP espouses values similar to those of the NLP.

Industrial and Farmers Fatherland Party – Governing Coalition

The IFFP is part of the current governing coalition alongside the NLP and the AFP. It is considered a far-right party by outside observers insomuch as it adopts a staunchly nationalist, sometimes stridently so, nationalist agenda. The party's appeal to the titular ethnic group of Northaria, the Northarians, has proven more attractive to voters in recent years. The success of the parties is arguably due to the mainstream parties' efforts to incorporate nationalist rhetoric to maximize their share of the vote against the growing bloc of ethnic Russian voters, which have generally supported the Rodina Patriotic Party (RPP). The IFFP emphasizes the uniqueness of Northarian culture and values (meaning it is not European); anti-gay; and blames the EU for its high inflation rates and lackluster economy. The main platform of the party is the protection of Northaria's sovereignty against encroachment from the EU or immigrants, particularly non-European one.

Rodina Patriotic Party

The RPP is a political party representing the ethnic Russian minority in Northaria. Most of its policy platforms revolve around addressing the perceived political, social, and economic inequalities many Russians in Northaria encounter. Since the RPP's founding in 2004, it has advocated for granting citizenship to all Russians born in the country and their children. In 2010, the RPP launched an initiative to have the issue of Russian citizenship put before voters, but was ultimately unsuccessful. Moreover, the RPP has fought to have Russian named an official language of government alongside Northarian in every municipality where 15% of the population is Russian speakers. Politically, the RPP is in favor of strengthening ties with Russia. It is also the only party represented in the Majvendi that openly opposes Northaria's membership in NATO and the EU. In the aftermath of Russia's annexation of Crimea, the RPP signed an agreement with the Crimean political party Russian Unity that spearheaded the political effort to secede from the Ukraine. The cooperation agreement was meant to strengthen the Russian world.

Analysis of Political Dynamics in Northaria

    a. The parliamentary construct of the Northarian constitution is adequate in managing the political economy of the country. It does, however, lend itself to the formation and progress of fringe, extremist parties. These can wield disproportionate influence when brought into a coalition government.

b. Given that the threat of Russian aggression is real and that ethnic heterogeneity in Northaria will be perpetual, the government and political leadership of the various parties must become accustomed to focusing on enhancing national integration, in addition to pursuing their other agendas. Russian propaganda aims at increasing social and political cleavages as a precursor to intervention in various forms. To forestall that, the Northarian political system must incline toward integration. A wide variety of methods may contribute to this end.

(1) Government leadership should examine the trends in education that Northarian politicians—both current and future—follow. With international assistance, Northaria should shape secondary education, undergraduate education, and post-graduate education toward liberal arts that encourage and normalize ethnic integration.

(2) The Northarian government, especially the Ministry of Education, should look abroad for undergraduate and post-graduate education opportunities and scholarships aimed at broadening the experience of potential political and business leaders.

(3) Government leaders should highlight successful legislation that features broad-based compromise and that cross-cuts ethnic cleavages. Political leaders should be influenced toward and rewarded for achieving cross-ethnic compromises. Conversely, leaders should address themselves to obvious incidents of polarization and seek to minimize them.

(4) The government should publish annual reports that detail the funding sources for the political parties that constitute the Majvendi. The goal is to expose disproportionate influence from abroad.

(5) The government and people of Northaria should consider constitutional amendments or laws that provide for a multi-ethnic cabinet.

c. The problem of corruption among political leaders is potentially catastrophic, because it makes the government vulnerable to Russian propaganda and alienates leaders from the people.

(1) Anti-corruption measures should include education, official scrutiny, exposure to media and watchdog groups, and an amnesty program that facilitates rapid transition to proper function.

(2) The Majvendi should also consider term limits for members as an anti-corruption measure.

Military

The Northarian National Armed Forces includes the Land Forces, the Navy (includes Coast Guard), the Air Force, and the Home Guard. There is currently no conscription, but the issue has been raised in the Majvendi every year for the past three years. Military expenditures have climbed to .91% of the budget last year (2015), and lawmakers seem well disposed to increasing expenditures.

Northarian land forces include an active force of approximately 5000 officers and men organized into two infantry battalions (one mechanized, one motorized), plus support personnel (artillery, logistics, signal, cyber, engineers, air defense, medical). There is also a reserve of approximately 11,000 that can be called in a national emergency or war. The Home Guard includes another 8000 soldiers (both active and reserve) organized into district defense battalions.

Pursuant to its membership in NATO since 2004, Northaria is developing special forces units in cooperation with the U.S. Army. Their mission focus is on stay-behind operations in the event of invasion.

Analysis of Military Dynamics in Northaria

a. Northaria should press for either focused recruiting program, some form of conscription, or an incentive program that provides substantial advantages to enlistees. Both military and nonmilitary programs may be considered. The goal is to provide an integrating shared socialization experience for all citizens of Northaria.

b. Northaria's special forces training program should be expanded with routine joint training. The goal should be improvement in defending against irregular and hybrid warfare. Northaria can work with potential sponsor states who share the same strategic goals to acquire defense articles, military education and training, and other defense-related services to support building Northaria's capacity defend its interests.[2]

(1) Northaria should plan, prepare for, and train for stay-behind operations. This includes detailed planning for logistical support, communications, medical support, and contingency planning.

(2) A key area for training and equipment is the need to maintain communications with the international community and leaders in exile during an invasion.

c. The Home Guard should focus their efforts on the threat of irregular forces infiltrating the country—including Russian spetsnaz agents, militias, and gangs. Crowd control techniques should be a major training objective. In addition, the Home Guard should train in protecting key government buildings, telecommunication centers, ports, transportation hubs, and military installations, because these are typically key targets for Russian infiltrators.[5]

d. The Ministry of Defense should conduct a thorough review of ports, roads, rails, and other critical infrastructure in coordination with NATO planners to ensure the rapid introduction of NATO combat power in the event of war. Ports must be capable of requisite throughput to facilitate wholesale logistics, and road networks must be able to handle anticipated retail logistics. The government should request routine NATO exercises to test infrastructure.

The national security threat against Northaria includes a spectrum of strategic options that the Russian Federation might pursue, escalating or de-escalating as the situation evolves. Under certain conditions, these could include outright invasion and occupation before NATO could deter or respond. The following section examines in greater specificity the role that stay-behind militias must play in the face of Russian aggression.

Northaria and American Unconventional Warfare

The United States Department of Defense definition of unconventional warfare (UW): activities to enable a resistance movement or insurgency to coerce, disrupt, or overthrow a government or occupying power by operating with an underground, auxiliary, or guerrilla force in a denied area.

It is most often considered as a strategic option when a nation has been taken over by a foreign aggressor or by a tyrannical government through a coup. In such cases, the U.S. and her partners may conduct UW to assist indigenous resisters in disrupting, coercing, or overthrowing the hostile government. According to U.S. doctrine, a UW campaign unfolds across seven phases. In this section we will examine how Northarian resisters could best cooperate with American forces conducting UW.

Phase I: Preparation

As the government of the Republic of Northaria begins to perceive a threat against their sovereignty, the environment will shift—sometimes gradually, sometimes rapidly—from steady state (i.e., routine peacetime political competition) into a pre-conflict state. During this period, the government may request assistance from the U.S. or other NATO allies. In American parlance, such aid could come in the form of "Foreign Internal Defense" (FID). FID is defined as participation by civilian and military agencies of a government in any of the action programs taken by another government to free and protect its society from subversion, lawlessness, insurgency, terrorism, and other threats to its security. If the threat to Northaria develops primarily as an internal insurgency, then U.S. aid may come in the form of "Counter-Insurgency"—i.e., comprehensive civilian and military efforts taken to defeat an insurgency and to address any core grievances. Finally, the government of Northaria might request assistance against a growing terrorism threat. In such instances, the U.S. would provide aid in the form of "Counterterrorism" (CT).

Unconventional Warfare, on the other hand, would commence after the legitimate government of Northaria has been overthrown in a coup or civil conflict, or when a foreign power invades, conquers, and occupies (perhaps even annexes) the country. In such circumstances, a portion of the Northarian population will seek to resist the new government with a view to overthrowing it and regaining independence. Assistance from the U.S. in this instance would come in the form of unconventional warfare.

The first phase of UW is "Preparation," during which the resistance and external sponsors conduct psychological preparation to unify the population against the established government or occupying power and prepare population to accept support. The first step in developing a UW campaign is the "intelligence preparation of the environment" (IPOE). During IPOE, American planners will gather extensive, comprehensive information about the resistance forces, physical geography, government forces, government strengths and weaknesses, social and economic conditions, cultural and religious context, and other relevant factors. UW planners must thoroughly understand the size, reach, and capabilities of the resistance, and determine whether the movement includes competing or cooperating factions. The indigenous resisters' level of training, logistical capabilities and shortfalls, and their political affiliations are key factors in understanding the potential for effective UW.

To improve its deterrence and defense capability, Northaria should prepare ahead of time to provide potential sponsoring states with this information. The details concerning geography, climate, infrastructure, and so forth should be on file for rapid retrieval when needed. Resisters need to strive for unity of command or at least unity of effort as soon as possible. They must establish redundant communications with sponsoring powers and continue to feed information concerning the enemy government's dispositions, capabilities, vulnerabilities, and operations. The more Northarian resisters can direct relevant information to UW planners, the better those sponsors can provide effective assistance. At the same time, Northarian planners must be aware of the security risks in gathering and keeping information on potential resistance. Russian intelligence will make that information a high priority, and elements within the government that might be friendly to the Kremlin could compromise resistance operations. For this reason, it is a good idea not to keep records of resistance networks and personnel.

In addition to preparing for UW operations, resisters likewise need to prepare for the post-conflict transition as well as how to sustain resolve over a possibly long and challenging campaign. They need to dust off or develop plans for transition to governance and a return to steady-state conditions. This includes thinking about how to reestablish governance at the national and local levels, how to rid the country of aggressors, and how to develop effective policy regarding amnesty and reintegrating former enemy factions. A poorly planned and led insurgency can degenerate into bloody civil war if not managed properly. In operations designed to achieve the status quo ante bellum, the objective of the insurgency is to overthrow the hostile government *and replace it with legitimate government.*

Phase II—Initial Contact

As part of the political decision making process in which the U.S. decides whether to render assistance to the Northarian resistance, Special Forces personnel, in conjunction with their interagency partners, will conduct the initial contact with representatives of the resistance. The purpose of this phase is for the sponsoring state to determine whether the resistance is viable and that its objectives and operations would be compatible with U.S. objectives. Conditions that are favorable for potential sponsor state support include a population that is open to participating in resistance activities, leaders who have goals similar to the sponsor state and support conducting activities in a humanitarian manner, and an adversary regime that lacks legitimacy and does not have effective control of the people or territory. To conduct initial contact, U.S. Special Forces may use a "pilot team," and they may arrange for representatives of the resistance to exfiltrate from the area of operation to meet with the pilot team. Obviously, the more Northarian planners can prepare ahead of time for such contact, the better they will execute when needed.

Optimal initial contact facilitates U.S. personnel to plan and coordinate support. This includes developing nonstandard logistics and assessing how best to build capacity within irregular forces in Northaria. During initial contact, planners also examine how best to provide Civil Affairs and information support to the resistance. This initial planning merges with the preparation phase in helping the UW forces to understand and dominate the land, maritime, air, and cyber domains in Northaria. Reconnaissance and security operations are ongoing through this phase and continue through all phases of UW.

Northarian war planners and government officials must plan ahead of time to facilitate initial contact. This is best achieved by developing relationships during peacetime and exercising initial contact activities. One key vulnerability during initial contact is the danger of Russian (or other) intelligence agencies infiltrating the resistance. The Cold War demonstrated that Russia excels at such measures, and both Northarian officials and American UW planners must take steps to properly vet representatives of the resistance. It is important for resistance elements and potential state sponsors to establish screening mechanisms and counterintelligence procedures to prevent infiltration of the resistance. The optimal approach would be that potential resistance leaders already have relationships with key figures of the Northarian government-in-exile.

Phase III—Infiltration

During this phase, the U.S. pilot team will infiltrate into the area of operations and continue the feasibility assessment on behalf of the U.S. government. If conditions appear favorable, a presidential finding may then call for a UW campaign to commence. The pilot team will then arrange the infiltration of follow-on forces. The teams likewise infiltrate supplies to increase resistance capability. As Special Forces team link-up with their respective counterparts among the resistance, they continue to develop the situation and communicate their updated assessment of their areas of responsibility.

Northaria can best support the infiltration phase by developing secure infiltration routes and "rat lines" for urgent escapes as necessary. Northarian planners, in coordination with U.S. planners, can outline key areas of the country in which resistance networks would likely develop. Members of the resistance can also communicate intelligence on government defenses against infiltration—radar sites, detection grids, etc. The key to success is continued secure communications.

Phase IV—Organization

Once U.S. Special Forces and Military Information Support Operations (MISO) teams have linked up with their resistance counterparts, the next step is to organize. This phase includes rapport-building between Americans and Northarians, as well as among the resistance organizations. Both the SF leaders and the leaders of the resistance communicate their expectations and objectives. The goal is to thoroughly align the two countries' strategic objectives, so that they are not working at cross-purposes.

Also during this phase, the resistance organizes its infrastructure to reinforce resiliency for the ensuing campaign. Concepts of logistical support, communications, movement, evacuation of casualties, etc. are worked out in detail among resistance groups and their American supporters.

Northaria can best assist the organization phase by eliminating or limiting factionalism within the resistance. Historically, eastern European countries during wartime have given rise to resistance movements within their countries that are at war with each other as well as the occupying power. Instead, Northarian political leadership must strive to create political, social, and cultural unity so that in an emergency, resistance groups are predisposed to work in the same direction with minimal conflict.

The side that controls or gains the sympathies of the population will have an advantage, improving access to information, recruits, and other resources, while denying these to its antagonists. For this reason, the resistance must develop political goals that have the broadest appeal and outline a compelling narrative that will help mobilize the population in collective action. Access to mobilization forums-- such as houses of worship, refugee and displaced-persons' camps, labor union assemblies, schools, and professional associations—is often of paramount importance to the resistance. These venues allow resistance leaders to communicate with potential followers. Virtual forums may be an option, but many adversaries will be capable of cyber surveillance and forensics.[6]

Phase V—Buildup

The purpose of the buildup phase is to grow the insurgency in numbers, capability, resilience, and support. Northarian resistance leaders will attempt to securely expand the network both in the countryside and in the cities (through underground operations). This expansion includes recruiting more auxiliaries as well. Likewise, the buildup phase includes negotiation to integrate splinter groups and others into the effort. Logistics, training, and reconnaissance and security operations occur throughout the buildup phase. The development of a counter-intelligence program within the resistance is necessary to prevent or minimize adversary infiltration. The use of biometric identification and surveillance could allow the USG and the resistance leaders to keep track of individual group members. The creation of a database containing resistance members would provide a useful accountability tool but also present a major security challenge.

Vetting and screening mechanisms, potentially relying on the recommendation or referral of trusted local partners (e.g., foreign intelligence services, clan leaders, former government officials, etc.) could help prevent adversary infiltration.

Northaria can best prepare for the buildup phase through programs that encourage proficiency in marksmanship, navigation, communications, logistics, and other skill sets needed during insurgent operations. A population that includes hearty outdoorsmen familiar with the countryside reduces the training burden on U.S. Special Forces and other resistance leaders. As above, visionary Northarian leadership can act during peacetime to facilitate the buildup phase. Achieving political and social integration, encouraging patriotism, and isolating or neutralizing dangerous factions that might cooperate with Russian aggression are key to success in the buildup phase.

Phase VI—Employment

The employment phase features actual insurgent operations in pursuit of strategic objectives. In the case of Russian occupation, the objective would be to cause or assist in the ejection of Russian forces and their proxies from the country. Insurgent tactics, including raids, skirmishes, sabotage, subversion, and so forth vary according to the operational plan. That plan may include the introduction of U.S. or NATO forces into the country. In that case, the insurgency can be invaluable in disrupting enemy defenses, reconnoitering enemy positions, and facilitating allied tactical operations.

It is critically important that during the employment phase, resistance leaders maintain a firm grip on insurgent groups and keep their operations aligned with the strategic plan. Economy of force is crucial in UW and insurgency, and it is therefore of paramount importance that neither efforts nor personnel be wasted on fruitless endeavors. Redundant and continuous communications among resistance leaders, the government-in-exile, and supporting nations are vital throughout the employment phase.

The resistance must continue to focus not just on tactical operations, but also on maintaining the proper narrative among the population. The struggle for legitimacy is of primary importance during an insurgency. This often requires leaders with vision, self-control, and political savvy. MISO teams can work with the public component to reinforce the insurgent narrative. As territory is liberated, the resistance provides population protection and support.

Phase VII—Transition

With the defeat of the hostile government, the focus of the insurgency changes to transition. This phase is crucial in achieving strategic outcomes. Resistance leaders must turn their attention to protecting and nurturing the new government. This requires the neutralization, integration, or elimination of former regime elements or other groups hostile to the insurgents and the new government. The transition phase must also include plans for demobilizing guerrillas and possibly offering amnesty to selected enemy groups. Messaging during this phase aims at reinforcing the government and building its legitimacy among the population.

Northaria can best prepare for such a transition by strengthening the public's confidence in government and educating the population concerning the country's traditions of democratic, limited government and respect for human rights. As before, MISO teams can be of great value, and media outlets—radio and TV stations, newspapers, internet news services, etc.—become part of the effort to restore peace and democracy. A preplanned demobilization of armed components of the insurgency can smooth the transition to civilian control and prevent the use of violence to affect peacetime political competition.

The Lesson of Gladio

The national security threat against the Republic of Northaria unfolds along a spectrum of conflict that stretches from outright invasion and occupation on the one hand, to irregular warfare, illegal and quasi-legal activities aimed at overthrowing or disrupting legitimate governance on the other. Given the nature of Russia's "New Generation Warfare" that capitalizes on non-kinetic factors and the full integration of political, economic, financial, cultural, social, and religious factors, the spectrum of conflict merges with routine political competition within Northaria. The line between legal political activities and Russian political warfare is ill-defined and hard to detect.

In general, however, Northaria's preparation for defending against Russian aggression must balance national security with the need to protect the country's constitution, law and order, and human rights. In the historical case of Operation Gladio, the fundamental mistake of those involved was the gradual development of the so-called "twofold purpose." As the prospect of Soviet invasion of Central and Western Europe retreated, the stay-behind networks became vulnerable to manipulation aimed at changing their purpose. As communist parties in various countries began to grow, network personnel began to assume the illegal and unconstitutional mission of interfering with political competition. The resulting terror, criminality, and subversion threatened the viability of the Western-style democracies.

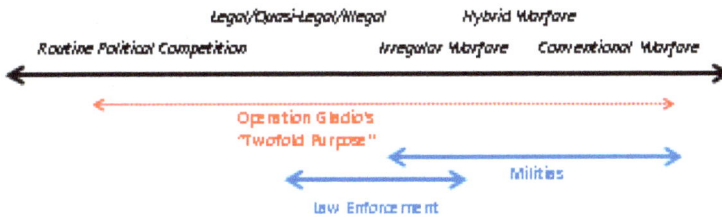

- The objective for the training and preparation of militias is for them to be ready once an enemy employs irregular warfare.
- Militias must have oversight and training to avoid the mistakes of Operation Gladio's "twofold purpose," which saw the use of militias to influence internal political competition.
- Triggers for the use of militias include the appearance of Russian proxies, unidentified military personnel, fifth columnists, paramilitary groups, terrorists, armed insurrection, or insurgents.

In contrast to those missteps, the modern militia system within Northaria must be trained, equipped, led, supervised, and vetted with redundancy to ensure that they are used only when Russian irregular warfare commences within the country. The trigger for such use would be the confirmed appearance of Russian proxies, unidentified military personnel (e.g., the "Little Green Men" who appeared in Crimea and eastern Ukraine), armed insurrection, illegal paramilitary groups (e.g., the "Night Wolves"), terrorists, or insurgents. Plans and operations for militias should be activated at the discretion of the national government in response to these triggers.

One lesson learned from the partisan movements of World War II and the Cold War was the necessity for paramilitary groups to maintain redundant communications with supporting external powers through more than intelligence or military channels. Instead, groups should seek continued contact with heads of state, a legitimate Northarian government-in-exile, key international organizations (e.g., UN, NATO) and, as appropriate, the media, including the internet. These redundant links to outside powers can better ensure that Northarian partisans' strategic objectives are aligned and coordinated with supporting powers—as did not happen during the Cold War.

The other salient lesson learned from that period was the danger of Russian infiltration of militias and partisan groups. Russian intelligence agencies excel at cultivating agents abroad and penetrating security measures. Indeed, partisan operations in the Baltic States from 1945-1956 were thoroughly infiltrated and disrupted by the Soviet KGB and its proxies. The safe assumption to make is that in the future Russian intelligence will infiltrate Northarian militias to some degree.

Economics

Structure of the Economy

For its small size, Northaria has a relatively diversified, resilient economy that has nonetheless undergone severe shocks in the past decade. Most of the country's gross domestic product (GDP), about 62%, is derived from the service sector, most of which is located in the major urban areas. The service sectors include telecommunications; food and accommodation; finance; education; and health care. Much like Estonia, Northaria has an advanced telecommunications industry for its level of development. The contribution of industry is high as well, contributing 28% of GDP. Agriculture, however, contributes only a small amount at 10%.

2008 Recession and Austerity Measures

In the early 2000s, Northaria was among the fasted growing economies in the EU. The peak of economic growth occurred in 2005 – 2007, when GDP growth reached nearly 10% each year. Economic growth was buoyed by foreign direct investment, particularly in real estate and construction, which reached as high as 25% of the country's GDP. However, the country experienced a crushing economic blow during the 2008 global recession. Northaria's GDP growth crashed by almost 18% in 2009, one of the sharpest GDP decreases in the world. The crash coincided with the collapse of the country's largest bank, Purity Holding Company. The country's Prime Minister (PM), Anu Lehtola, requested an 8-million-euro bailout from the International Monetary Fund (IMF) and the EU. As she did during other economic crises driven by the recession, IMF president Christine Lagarde demanded PM Lehtola implement austerity measures to ensure fiscal discipline.

The recession had formidable economic and social repercussions for the country. The Baltic States, including Northaria, experienced the collapse of a housing bubble in major urban areas precipitated by cheap credit available from Scandinavian banks. By 2007, housing prices had increased nearly 95% since the early 2000s compared with a more modest increase of 11.8% in the EU during the same period. Midway through 2008, housing prices had collapsed, plunging by over 60% before beginning to recover in 2010. During the recession, unemployment figures also skyrocketed. In 2007, the unemployment rate stood at 8.5%. By 2009, that figure rose to a high of nearly 20%, with youth experiencing unemployment at almost 30%.

Northaria has undergone a limited recovery since the devastating crash of 2008. Although the country has not recovered its pre-recession double-digit growth rates in past years, it recorded a 1.5% annual GDP growth rate in 2015. Although the GDP growth rates are promising, the increases are small gains compared to the enormous 2008 losses. Northaria still has not recovered its pre-2008 GDP. In 2012, Northaria qualified to enter the Eurozone, the second Baltic State to do so after Estonia. Moreover, unemployment has recovered significantly since the recession. In 2015, the unemployment rate stood at 10%.

However, plummeting unemployment rates are driven in part by emigration as the country hemorrhages its skilled workforce to Germany, the UK, and Scandinavian countries in search of employment and higher wages. An estimated 3.7% population drop is attributed to employment emigration during this time, a trend that disproportionately impacted rural areas as opposed to more affluent urban areas. The demographic shifts are compounded by the crushingly low birth rate in the country. The IFFP, with some support in the NLP, has called for more generous maternity leave to offset the low birth rates among ethnic Northarians. Some leaders in the IFFP have also called on the EU to compensate the country for its human capital losses. The calls have included demands that EU countries with significant numbers of Northarians teach the children in their native language at school. The high rates of emigration have further contributed to growing Euroskepticism as the freedom of movement within the EU, once its greatest selling point, has hamstrung economic growth and demographic viability in the country.

Northaria's economic policies in the aftermath of recession relied on austerity measures, like much of Europe, in contrast to the U.S. that focused on fiscal stimulus as the primary recovery mechanism. Not only had the country received a loan from the IMF, which required austerity measures, but Northaria also continued pursuing integration into the Eurozone, which also necessitated fiscal discipline. This means ensuring that public debt to GDP ratios remained low, requiring devastating cuts to social benefit programs. Austerity measures included slashing public sectors wages on an average of 30% and cutting jobs by nearly a third. Other entitlement programs were treated similarly, some of which have exacerbated the demographic difficulties facing the country. The government reduced childcare and paternity leave by 40%; unemployment payments were cut from nine months to six months; and pensions were cut by 10%. Moreover, government spending also decreased considerably on transportation infrastructure, health care, and education. Spending on healthcare was particularly hard hit, leading to a reduction in emergency services by nearly 20% and the shuttering of hospitals and clinics throughout the country. Human development indicators, including crude death rates and maternal death rates, all increased during the period of austerity. While Northaria's economy has improved, its infrastructure is still limping to recover from the austerity cuts.

Ethnic Russians in the Local Economy

In Northaria, Russians are more likely to hold lower wage jobs than their Northarian counterparts. Russians anecdotally report that they have difficulty finding work without fluency in the Northarian language. Some report the difficulties despite having university degrees. Many feel that in order to be competitive in the job market, fluency in Northarian is a necessity. However, fluency in Northarian, as well as Russian, is generally regarded as a competitive skill set. It is not uncommon to encounter merchants in Russian neighborhood enclaves in which only speak Russian. A division is notable between the higher-end merchants that cater to Northarians and those that cater to ethnic Russians. Russian merchants are separated, both in terms of structure and signage, from mainstream merchants.

Analysis of Economic Dynamics in Northaria

a. The government should consider legislation that mandates businesses providing free instruction in the Northaria language for its employees. Likewise, tax incentives for ethnic diversity within work forces should be offered.

b. Infrastructure repair and maintenance should be encouraged with tax incentives and favorable regulations. The government should encourage construction companies with generous contracts to restore infrastructure, especially in rural areas using local labor.

c. The requirement for austerity and the resistance against it should be addressed by the government teaming with industry leaders to replace government-provided benefits with employee-provided benefits. As above, companies that comply should gain tax incentives.

Society/Demographics

The population of Northaria is currently at 1.8 million. Northarians are the titular ethnic group of Northaria and are the largest ethnic group in the country. Linguistically and culturally, Northarians are part of the Finno-Ugric peoples, which also includes the Finnish, Estonians, Hungarians, and others. Northaria, Finland, Estonia, and Hungary, totaling about 21 million of the existing 25 million Finno-Ugric peoples, are the only nation-states representing Finno-Ugric peoples. Finno-Ugric populations are generally enclaves in much larger Indo-European populations, such as the 17 separate Finno-Ugric peoples that live in Russia. The native Northarian tongue, like Estonian, is a Finno-Ugric language, although native speakers of each language are not fully intelligible to one another due to dialectical differences. Other Finno-Ugric languages are less intelligible, with only some root words in common available to distinguish linguistic kinship. Northarians identify much more closely with Nordic than Slavic peoples.

Ethnicity	Population	Percentage
Northarian	1,134,000	63%
Russian	486,000	27%
Baltic (Estonian, Latvian, Lithuanian)	90,000	5%
Other	90,000	5%

Like most Finno-Ugric peoples, Northaria is almost entirely Christian. Ethnic Northarians are overwhelmingly Lutheran, around 98% of the population. However, the religious designation is a misnomer. While most can identify a distinct religious tradition to which they belong, around 80% of the population consider themselves irreligious. Church attendance in Northaria is among the lowest in the EU. Ethnic Russians, by contrast, are almost exclusively Russian Orthodox. There are a small number of Jews in Northaria, numbering around 3,000. Similar to Estonia, ethnic Northarians initiated few recorded pogroms against the Jewish population during the Nazi German occupation of the early 1940s. In nearby Latvia and Lithuania, the Jewish communities were less fortunate and suffered devastating pogroms that succeeded largely due to the collaboration of local Christian populations with the Nazi occupiers.

Geographic Distribution and Integration of Ethnic Minorities

One of the most important legacies of the Soviet era is the demographic shifts resulting from Soviet migration policies. In the Stalin era, before, during, and after World War II, Soviet policies displaced tens of thousands of Northarians from their homes. Many were sent to Siberia or resettled in others areas. Still others, particularly those with less desirable political views or from less desirable classes, were sent to prison camps where most of them died. Stalin also used the deportations to implement agricultural collectivization, an effort to consolidate individual farm holdings into collective enterprises that would increase the country's food supply. Historians estimate that the deportations, which continued until the de-Stalinization efforts of Nikita Khrushchev in 1956, lost at least 5% of its Northarian population. Others estimate the numbers are as high as 10% of the population. After 1956, some Northarians returned home, but most did not. In 1989, the Supreme Soviet of the USSR acknowledged the deportations as unlawful and criminal.

The second set of Soviet migration policies encouraged the migration of ethnic Russians to Northaria. After the Soviet occupation of Northaria, ethnic Russians were encouraged to migrate to the country to help industrialize its largely agricultural economy. The influx of migrants, alongside the forced deportations, contributed to significant demographic alterations in the country. While Northarians comprised a dominant majority in the country before the bulk of Soviet policies came into effect, by the mid-1970s the percentage of Northarians in the country dropped to just over 55%. During the tumultuous period surrounding the dissolution of the Soviet Union, ethnic Russian migration out of Northaria contributed to a rise in the share of ethnic Northarians in the country. Overall, the percentage of Northarians as a portion of the population dropped from 64% at the end of the Second World War to just over 60% at the time of independence. Many nationalist-leaning Northarians regarded the shift as an existential threat to their culture.

Currently, ethnic Russians and Northarians are moderately segregated spatially, linguistically, and socially. The Russian populations are concentrated in two primary regions. The highest concentration of Russians is in the city of Skaiyae in eastern Northarian. A city of about 75,000, Skaiyae's population is over 70% Russian. In addition, there is also a significant population in Northaria's capital, Mestauskal, comprising about 37% of the population in the city. Besides the Russian enclaves of Mestauskalio and Skaiye, minor neighborhood enclaves are scattered throughout urban areas in Northaria. The rural areas of the country are inhabited almost entirely by ethnic Northarians.

However, the geographic distribution of ethnic Russians and Northarians are not the only pertinent factor separating the two groups. After Northarian, the second most widely spoken language is Russian. The last census reported that Russian is the native language for around 38% of the population with the Northarian the native language for 57% of the population. However, around 20% reported Northarian as speaking a second language. Most ethnic Russians are educated in minority schools that offer instruction in Russian. In 2016, however, a bill sponsored by the IFFP, and

Figure A-4: Protests over Language Discrimination[3]

passed by the Majvendi with support from the NLP and the AFP, requires that 60% of all instruction in educational settings be conducted in Northarian. The move prompted protests, some say organized by the youth wing of the RPP, outside the Majvendi. Protestors held signs that seemed to threaten a merger with Russia if language discrimination continued, alluding to similar outcomes in the Ukraine.

Post-Independence Citizenship

Among the new country's most controversial policies in the early days of independence was the question of citizenship for its minority population. During the Soviet era, the CPSU undertook central economic planning efforts to rebuild and modernize the Baltic States. Over the course of several decades, Russian labor migrants flooded Latvia, Estonia, and Northaria, although less so Lithuania which maintained a healthy workforce. In the case of Northaria, the CPSU intended the Russian migrants to industrialize the predominantly agricultural economy of Northaria. By the 1970s, Russians comprised almost 36% of Northaria, leading to fears that Northarians would soon be without a homeland due to the demographic pressures. Once Northaria declared independence, it was not immediately clear how citizenship would be determined. In 1991, Russians made up around 30% of the population, although the numbers have dwindled due to emigration and naturalization. Eventually, it was decided that only Russian families that could demonstrate they migrated to Northaria prior to its absorption into the USSR in 1940 were eligible for citizenship. This proved difficult, if not impossible, for many Russian families. As a result, many Russian residents within Northaria were effectively disenfranchised from their country of residence. Since that time, the Majvendi has adopted slightly more lenient policies regarding citizenship, but ethnic Russians are still required to qualify for citizenship based on Northarian language tests. The tests remain a significant barrier to citizenship for Russian residents in Northaria, particularly older generations that have had limited exposure to the Northarian language in secluded ethnic enclaves in the country's major urban areas.

While the initial citizenship laws were highly exclusive of Russians, the EU pressured Northarian leaders to introduce laxer laws. In 1996 and again in 2003, Northaria relaxed citizenship laws, dropping some requirements or making citizenship tests easier and more accessible. As a result, during this period, thousands of ethnic Russians applied for and received citizenship. Although the citizenships laws in Northaria have relaxed since independence, some ethnic Russians have not applied for citizenship. Russians that are eligible to apply but have not are regarded with suspicion by Northarians, particularly among constituents of the IFFP. Several IFFP MPs have stated that such individuals are not to be trusted and should be regarded as agents of Moscow since they are deliberately avoiding naturalizing. However, there are some legitimate reasons why ethnic Russians may not have applied for citizenship. Russia itself gives preferential treatment to ethnic Russians that are not citizens of the Baltic State in which they reside. For instance, Russia does not require noncitizens to apply for visas to cross the border to visit relatives or to find work. Some are also eligible for pensions paid by the Russian state despite residing in a foreign country.

Rising Nationalism in Northaria

There are several factors that have prompted the emergence and electoral wins of the far-right parties: shifts in socio-economic and socio-cultural cleavages. Difficulties in the country's economic development have left opportunities for far-right wing parties to emphasize the population's socio-cultural cleavages, especially as regards anti-immigration rhetoric and a return to the titular group's traditional values.

The slow creep of globalization brought on by integration in the EU is increasingly viewed as a threat to Northaria's unique culture. Membership within the EU brought a great deal of foreign investment and political influence from Western countries. Non-state international organizations, like the Soros Foundation, also provide resources and support for liberal democracy, anti-corruption, and non-state civic actors. The Soros Foundation is among the largest Western NGOs operating in Northaria. The supporters of the IFFP identify mainstream liberals as "sorosisti" after the Soros Foundation. They charge the sorosisti with being pro-gay and secular while trying to rouse resentment against the Northarian political and economic elites they charge with corruption.

Northaria's far right-wing groups have also benefitted from public sentiments of mistrust for political parties. As mainstream parties lose some voters due to mistrust issues, the right-wing parties are able to attract them. Moreover, the mainstream parties are also moving close towards the political center so voters at the end of the far ends of the left-right spectrum have been without a party that voices their view. These are the voters that the far right is attracting. The inclusion of the IFFP in the governing coalition has also produced a snowball effect regarding the norms surrounding the inclusion of nationalist rhetoric among mainstream political elite. Support for far-right nationalism has quickly become the "new normal," and thus, a more legitimate, socially and politically acceptable position to adopt. As a result, even the mainstream is comfortable endorsing anti-immigrant policies while issuing sharp criticisms to EU leaders like Germany's Chancellor Angela Merkel that have supported immigration in the past. As refugees continue to pour into Europe from the destabilization in the Middle East and North Africa, it is likely that the nationalist trend in Northarian politics will continue or even worsen.

Far-right wing parties are supported by non-party organizations and subcultures that are characterized by verbal aggression, radical nationalism, and ethnic prejudices. Most of the subculture revolves around ties to similar groups in mainland Europe and are particularly active on the internet and in social media. In the past several years, existing far-right subcultures have been bolstered by the import of the Finnish vigilante group, the Soldiers of Odin (SoO), which has opened chapters as far away as the United States,

Figure A-5: Soldiers of Odin Participating in Anti-Migrant Protest[4]

even attending the Republican National Convention there in support of presidential candidate Donald Trump. The youth wing of the IFFP was among the first to implement the SoO patrol units that claim to be protecting Northarians, particularly women, against attacks from migrants and ethnic Russians. It is estimated that the SoO have managed to attract around 8,000 members in Northarian since the opening of the first chapter in 2015. The group claims to patrol major urban areas in 24-hour cycles to protect residents from threatening behavior and situations. Observers have accused the group of far-right extremism in connection with founder Mika Ranta who has been convicted of racially motivated violence in his home country. However, in interviews with media, the Northarian chapter of the SoO claims it has taken efforts to distance itself from white supremacist ideology and its supporters.

The rising nationalism and anti-immigration stances have resulted in concrete policy changes in the last year. The first is the government's statement that it would accept no further refugees despite the urging of Chancellor Merkel to expand its refugee program. Instead, the government has begun construction of a fence surrounding its border with Russia. Since late 2015, refugees have taken

Figure A-6: Fence Separating the Northarian and Russian Border to Prevent Entrance of Refugees[5]

advantage of the relatively porous border to gain access to Europe. Few of these refugees are likely to stay since Northaria has few social programs available and has harsh, cold weather.

The IFFP youth wing and the associated chapters of the SoO far-right nationalist perspective has coincided with a rise in anti-Russian sentiment. Protests, minor riots, and attacks against Russian interests in Northaria have seen an uptick in recent years, particularly after Russia's aggressive unconventional warfare efforts in the Ukraine. Since World War II, nationalism and anti-Soviet or anti-Russian sentiment have often coincided. After the Soviet occupation of the Baltic States in the Molotov-Ribbentrop pact, Germans invaded the Baltics to secure the territory for the Third Reich. During this

Figure A-7: Commemoration of Legionnaires as Veterans Lay Flowers at the Foot of the Liberation Statue

period, some 35,000 ethnic Northarians joined units of the German Waffen SS, although it is unclear to what extent the Northarians were volunteers or conscripts. Regardless, in collaboration with the Germans, Northarian soldiers fought against the Red Army to end Soviet occupation of the area. The Liberation Statue, located in Northaria's capital, commemorates the efforts of the Northarian soldiers to secure independence from Soviet rule. However, detractors of the Legionnaires, as the Northarian German collaborators are called, accuse Legionnaire supporters of being thinly disguised neo-Nazis, claiming that any glorification of the Legionnaires is also a glorification of the Nazis and the atrocities associated with the Holocaust. Legionnaire supporters deny that the soldiers embraced Nazi ideology and instead paint them as "Freedom Fighters" for the cause of Northarian independence. The IFFP, alongside its youth wing and the SoO, are public supporters of the Legionnaires. At the last commemoration in May 2016, the groups and their supporters held a candlelight vigil to mark the day. In the late evening hours, several hundred participants also marched to the Red Warrior statue, which commemorates the Soviet victory over Germany, and defaced the statue. Minor riots followed the acts of vandalism as anti-Legionnaire protestors and ethnic Russians converged on the spot as it was being reported on social media. The Northarian PM has threatened to ban the commemoration to avoid a further escalation of violence, but it is unknown whether he has the support at this time to do so.

In recent years, support for the Legionnaires has become more commonplace. The Northarian Minister of Defense publicly thanked the veterans for defense of their homeland during the annual commemoration of Legionnaire's Day in May 2016. His comments were met with harsh condemnation by leaders in the NLP, but the Minister faced few, if any, repercussions for his comments. The IFFP has also issued calls for the removal of the Russian statue, but the NLP has so far hesitated to remove it, aware that such a policy could tarnish its human rights record with the EU and provide more leverage for Russia to accuse the country of discriminating against its Russian residents. The RPP has accused the NLP and the IFFP of corruption and bias, saying the head of the country's governing coalition is complicit in the rise of anti-Russian sentiment and a recent spate of vandalism that has attacked Russian businesses, schools, and neighborhoods. Thus far, the damage of the attacks has been minor, but the

RPP has called on the NLP to issue harsher condemnations of the attacks and to increase security presence around Russian interests in the country.

Calendar Mobilization Days

Calendar mobilization days are commemorative holidays that serve as important moments for mobilization on behalf of an individual's or group's political beliefs. There are two important commemorative days in Northaria that have sparked confrontations between far-right nationalists and pro-Russian supporters.

- June 22 (Legionnaires Day): A day commemorating the Northarians drafted into the Waffen SS, to fight Communist Russia for Northarian independence. There are usually processions to the Liberation Statue, located in Northaria's capital. Russians tend to take offense at the overtly nationalist tone in the celebration. There are periodic attempts to take down the monument that spurs heated rhetoric and low-intensity violence among the opposing sides.

- May 9 (Soviet Victory Day): This day commemorates the Soviet victory over the Germans during World War II. Ethnic Russians widely celebrate the holiday. Generally, it includes some sort of processional or wreath laying at the statue of the Red Warrior. Some years, the parades to the statue are led by the Russian ambassador to Northaria. Nationalists, and supporters of the titular ethnic Northarians, take offense at the holiday since it is a celebration of the Soviet occupation of the country. The holiday, however, has not been an official national holiday since Northaria claimed independence from the USSR in 1991. Some

Northarians instead celebrate the defeat of Nazi Germany on May 8, the day the Allies accepted the surrender of Germany.

Analysis of Social Dynamics in Northaria

a. The observed increase of nationalism and ethnic tension in the country constitutes a potentially existential threat to Northaria.

b. The goal in managing the problem is to overcome ethnic tension through both integration and the creation of cleavages that cross-cut ethnicity. For example, membership in broad-based labor and trade unions can become more significant to an individual than his or her ethnic identity.

(1) Integration measures include education, sports programs, language programs, and media management.

(2) Cross-cutting cleavages can include multi-ethnic political parties, labor unions, trade unions, religious organizations, social clubs, etc.

c. The management of language issues is central to resolution of ethnic tensions. Typically, governmental attempts to restrict the use of a given language are problematic. Instead, the government should encourage the use of both Northarian and Russian in most of the population. The government should offer financial and tax incentives or other advantages to those who achieve bilingual proficiency.

d. Political leaders should seek to defuse tensions that arise during mobilization days. Instead of resisting such celebrations, leaders should guide the population and especially their political constituents in joining them. The government should seize the initiative on this matter and push legislation that widens the celebrations and encourages discussion concerning them. By honoring ideas and institutions that minorities hold dear, the government can encourage closer emotional and spiritual ties among the people.

e. The objective in managing social dynamics is to view the Russian compatriots in Northaria as an advantage, rather than as a disadvantage. By highlighting and facilitating their success, leaders can disarm the Kremlin and remove dangerous issues that could lead to intervention. The goal is for Northarian ethnic Russians to feel loyalty to the state as well as pride in their ethnicity. They should come to think of themselves as full-fledged members of Northarian society with interests and identities in matters other than ethnicity.

Information and Infrastructure

Ethnic Northarians and Russian minorities generally receive information from different media sources. The latter receives nearly all of its information from Russian-language media supported by Russian authorities. The former group receives information from a variety of sources in their language.

Transport Infrastructure

Northaria's transport infrastructure is adequate for a highly developed country in urban areas, but the quality suffers significantly in rural areas. Motorways, which cover about 150 km, are in need of further development, but generally, the country is well connected via its roadways. Local roads are about 42,000 km while the state roads are about 19,000 km. Quantity is not necessarily an issue, but the quality of Northaria's roads is problematic. Many kilometers of roads are in need of upgrading and repair. Moreover, in rural areas, around 40% of the existing road network is unpaved. Road quality is further impacted by harsh conditions during the long Nordic winters.

Figure A-8: Typical Unpaved Rural Road in Northaria.

Rail networks are of much less importance in transportation in Northaria than in the rest of the EU. The small country has only 1150 km of railway, making it one of the least dense rail networks in the EU. The significance of the rail networks for transportation has diminished considerably since independence as the Northarian population becomes increasingly motorized, relying on personal vehicles for most transportation needs. Passenger rail transport accounts for only about 5% of all passenger kilometers. The number of rail passengers has dropped nearly 50% since 1994. However, freight transport still maintains a dominant position. The payload distance of freight in Northaria, measured as tons per kilometer, has increased by 300% since 1994. Nearly half of freight transports are oil or oil products.

Public transportation in Northaria's capital, Mestauskal, and the surrounding areas is extensive and free to all residents. The transit is a mix of tram and trolley, with around 850 km length combined. There are 6 tram lines and 10 trolley lines in total. Since 2004, there have not been any additions to either system. While most of the transit system is operated publically, some portions of the bus transit are operated by private companies. Rural public transportation is almost exclusively provided by bus, but it lacks a structured, scheduled system. The private companies operating the buses have few tools available to determine market-driven demand, routes, and scheduling. Ridership has been decreasing across all public transportation modes as the individual systems age, experience workforce problems with an aging population, and struggle to provide regular, reliable service.

Energy Infrastructure

Northaria's energy infrastructure has improved significantly since its accession to the EU in 2003. The country relies primarily on shale oils (about 60%) for its energy needs, but also crude oil and natural, although to a much lesser extent. Northaria has the largest supply of oil shale in the Baltics, about 4,000 km^2, representing about 1.3 billion tons of usable oil shale resting 10 – 80m below the surface. Although there are numerous rivers in Northarian, the primarily flat terrain reduces the possibility of viable hydroelectric plants. Only several exist, providing a small percentage of the country's energy needs.

The electric grid is managed by a private company, Telerung, which is responsible for the planning, functioning, and managing of the extensive grid network and its international connections. Northaria's electric grid is connected to the EU electricity market through Nordlink 1 and Nordlink 2, connecting in Finland. In total, the grid is comprised of 5,946 km of power transmission lines, including 330 Kv, 220 Kv, 110 Kv, 35 Kv, and direct current lines. The electric grid also has 158 substations. Nordlink 1 and Nordlink 2 are undersea cables in the Gulf of Finland linking Finland and Northaria. The connectors were critical to ensuring Northaria's independence from Russia's energy market, a vulnerability that the Kremlin has used for political leverage in the past. During the installation, Russian naval ships repeatedly interrupted installation of the line, according to Swedish sources.

The electric grid infrastructure is sufficient to produce enough for internal electricity demand with a peak load of about 1,956 mw. It also exports electricity to Latvia and Estonia. The largest energy producer in the country is North Energia, which supplies nearly 90% of the electricity and heat for the country. Around 13 – 15 million tons of oil shale is delivered to North Energia's power production plant in Skaiyae via rail in 300-400 rail cars or up to 600 rail cars in winter. The shale is offloaded and sent to impact crushers via belt conveyors where it produces energy through a boiler and turbine system.

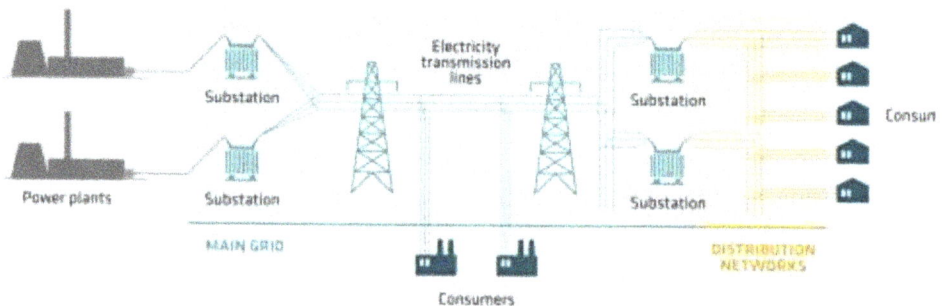

Figure A-9: Diagram of Northaria's Electric Grid Infrastructure

Northaria relies less on natural gas than shale oil. However, Finland and Northaria signed agreements to establish greater energy connectivity between Northaria, Finland, and Estonia. The private energy firm, Baltic Connector Oy, based in Finland, is constructing a subsea gas transmission pipeline to facilitate transfer of gas to Estonia and Northaria. Once finished in 2019, the pipeline will help end Northaria's isolation from global gas supplies and mitigate its reliance on Russian imports for its energy supplies. In 2016, Russian gas juggernaut Gazprom sold its 40% stake in North Energia. The sale was likely precipitated by Baltic efforts to reduce its dependence on Russian gas, such as through the Baltic Connector effort.

Maritime Infrastructure

Northaria's major ports are located on the Gulf of Riga, with easy access to the other Baltic States, Finland, Sweden, and Russia. Altogether, Northaria has 32 seaports and 9 inland ports. Northaria's capital, Mestauskal, operates 9 ports in its jurisdiction. All of the seaports are operated by publicly held companies at the municipal level. While privately owned ports are emerging in nearby Estonia, Northaria's remain in public hands for now. Nearly all of Northaria's sea ports, including all of those operated by the Mestauskal municipality, have intermodal transport that easily connects dry bulk and liquid petroleum products cargo by rail and sea. Around 70% of the cargo transited through the ports is liquid petroleum products, the rest being dry bulk. Russia is the heaviest user of Northaria's ports, using them to export oil to Western Europe.

Telecommunication Infrastructure

Northaria, alongside its neighbor Estonia, ranks among the world's most wired and telecommunication advanced countries in the world. The country has an astonishingly high rate of high-speed internet penetration and utilizes high levels of e-commerce and e-government services. Moreover, the country is consistently ranked among the highest in the world on freedom of information. However, Northaria's technological advancements have increased its vulnerability to cyberattacks and the government struggles with issues related to privacy of information and containing politically incendiary, xenophobic, and racist hate speech on its many social media platforms.

Northaria's telecommunication infrastructure was in a state of extreme disarray when it first gained independence from the USSR. In many regards, this worked in the country's favor since it did not have to adapt existing infrastructure to accommodate technological advancements. After independence, Northaria's legislative leadership, many under 40, recognized the importance of the expansion of telecommunication technology to sustainable economic and human development. As a result, Northarian political leadership made investments in this domain a top priority.

The country's leadership established the first internet connections in government facilities. Subsequently, academic facilities in Mestauskal and Skaiyae received connections as well. After the initial connections were established, the national telecommunications monopoly was privatized, becoming Snap Ühend, entering into partnerships with Norwegian and Swedish companies. Snap Ühend laid fiber-optic cable throughout the country capable of delivering fixed and mobile communication services. Working in concert with Northarian political leaders, the company also unveiled plans in 1996 to provide internet access and technology centers to every public school in Northaria. By 2000, Snap Ühend reached its goals. The result is a high level of computer and coding literacy among the country's young and up-and-coming labor force. Current levels of internet penetration are around 83%. Moreover, virtually all urban areas are covered by public access wifi connections, including inside hospitals, schools, hotels, gas stations, grocery stores, and all public facilities. Wireless access, based on Code-Division Multiple Access (CDMA) technology, also extends into rural areas as well, covering about 88% of Northarian territory. The service is priced at rates competitive with fixed broadband access. Municipalities, with funds from Mestauskal, have helped to subsidize deployments of the CDMA technology at the local level. Moreover, government regulations have helped to lower the costs of entry to the market, enabling a proliferation of local startups to help manage the workload. Levels of mobile subscriptions are high, reaching around 1.2 million, which indicates that many Northarians own multiple mobile devices. Four mobile operators provide Northaria with mobile service, including 3G, 3.5G, and 4G services, covering 98% of the territory. Internet users in Northaria use the services for a wide variety of applications, including search engines (93% of users), email (89% of users), local media, social-networking sites, Voice over Internet Protocol (VoIP), and instant messaging. The public broadcast channels also deliver its radio and television production services online, including news in real time.

Northaria has the largest public-key infrastructure (PKI) in Europe, second only to Estonia. The PKI is based on electronic certificates maintained on the national ID card database. Around 1.1 million PKI are currently in use, enabling electronic authentication and digital signing. Around 56% of PKIs have been used for these purposes. Northarian regulations make the digital signature have the same legal weight as a handwritten signature. The expansive PKI has allowed Northaria to migrate nearly all of its government services to a virtual e-government bureaucracy. Currently, all government functions are accessible to citizens, using the PKI, through the e-government portal. This also includes the payment of taxes and voting, both of which can be done online. In the last election, 58% of Northarians that voted did so online. Additionally, around 98% of Northarians paid their taxes online.

Cybersecurity in Northaria

Northaria's impressive gains in its cyber infrastructure have revealed some significant downsides. While electronic infrastructure is a force multiplier in political, social, and economic domains, it also exponentially increases vulnerability to disabling cyberattacks. Northaria, alongside Estonia, was the victim of the world's first cyber war in 2007 following criticisms of the countries' policies on remaining Soviet iconography in the Baltic States. The conflict began after a controversial decision to move the Soviet Unknown Soldier statue to a military cemetery from its original home in a central square of Estonia's capital city, Tallinn. The statue had long served as a rallying point for Russian and Soviet-related celebrations by the hundreds of thousands of ethnic Russians residing in the country. Ethnic Estonians and Russians rioted in the streets for days following the decision, leaving at least one participant dead, scores injured, and hundreds arrested. During the debacle over the Unknown Solider, the Northarian Prime Minister voiced staunch support for the decision to move the statue, even considering doing something similar with the commemorative Red Warrior statue in the capital city. Riots erupted in Northaria between Northarians and Russians, mirroring the confrontations occurring in Estonia.

The physical confrontations between the competing factions were only the opening act in the salvo. Several days after the riots in the streets began, hacktivists launched a two-phased attack against the Northarian and Estonian digital infrastructure. The first wave of the attack, which appeared to be led by amateurs, hacked into the websites of each government. The attackers posted an apology letter, ostensibly signed by the Northarian Prime Minister, apologizing for its support of Estonia. The letter included a picture of Estonian Prime Minister Andris Ansip emblazoned with a swastika. It denounced Estonia's policies as fascist and ultranationalist, promising to distance Northaria from the "ethnocratic" regime in Tallinn. Meanwhile, Russian-language web forums hosted both within Northarian and inside Russia provided detailed instructions on how to hamstring Northaria's digital infrastructure using distributed denial-of-service attacks. The attacks would overload Northaria's server capacity with unprecedented levels of traffic, effectively shutting down vital services. This second phase of the attack, funded through PayPal accounts, was successful in orchestrating a network of botnets, or zombie computers, that were directed to Northarian electronic portals. The networks included millions of computers spanning the globe. The attackers orchestrating this phase are widely regarded as more sophisticated, highly trained, and well funded than those participating in the first wave of the attacks. The successful attacks shut down the Northarian government's email system, public school email systems, and services for a wide range of bank and other commerce sites. The counterattack, spearheaded by Estonian and Israeli cyber experts, limited the potential secondary effects of the closure of vital systems. The denial-of-service was limited to a period of several days in most cases. The attacks led to the establishment of the NATO Cooperative Cyber Defence Centre of Excellence in Tallinn, Estonia.

After the attacks, Northaria implemented a series of measures designed to lessen the impact of another similar attack, increasing the country's resiliency against cyber warfare. The primary secondary effects of the attacks are the experience and motivation cyber experts in the region gained in dealing with the attack. Northaria created the Cyber Defense League, which is a volunteer organization that falls under the Ministry of Defense. Its purpose is to protect Northaria's cyberspace domain. Its members include civilian IT security specialists, other specialists in related fields, and youth with skills and interest in cybersecurity. Northaria has also instituted training in the public-school system to introduce all students to the basics of tech and code necessary to protect Northaria's highly digitized society. Like a regular military civil defense organization, the Cyber Defense League has regular training schedules on the weekend under a unified military command. The private sector is heavily represented in the Cyber Defense League, a vital component of cybersecurity since much of the country's infrastructure, as in the United States, is in private hands. This includes transportation, power, and the financial industries. Northaria is also considering conscripting for the cyber service if the need arises.

Media Consumption

Russians are among the world's most agile information operation actors in the world. In the Baltics alone, Moscow maintains a budget of nearly $300 million for its propaganda campaigns. Its efforts are supported by academic partners in major Russian universities. For the most part, ethnic Northarians and Russians consume media from separate sources, divided primarily by language. Newspapers, radio stations, and TV channels for each ethnic group is separated according to language, but also by content and perspectives on the Western world, especially as regards Western Europe, NATO, and the U.S. Among native Russian speakers, around 73% read newspapers and magazines regularly. This reading population has access to four Russia-languages newspapers and two Russian-language magazines, although the publications are expected to shrink in number alongside most print media. The primary Northarian paper, *Mestauskal Post*, also publishes a Russian-language edition twice each week with highlights of major stories. Radio is also a popular medium among Northarians. Around 70% of ethnic Northarians listen to the radio, although radio is slightly less popular among Russian speakers. In all, there are three radio stations that broadcast exclusively in Russian, although residents in the southeast of the country also have access to broadcasts from within Russia itself. The radio broadcasts of the Russian media project, Sputnik International, are also received by most Northarian households. Sputnik International is owned by Russian news agency, Rossiya Segodnya, but receives tight oversight by the Russian government. It is widely regarded as a Russian government propaganda outlet by the U.S. and the EU. Among TV stations, ethnic Northarians generally watch TV2, Kanava10, and NTV, all broadcast in Northarian. Russian speakers rely on the First Baltic Channel (PBK), NTV Mir, and RTR Planeta. All three of the stations are owned by Russian media firms and broadcast exclusively in Russian. PBK is the most popular channel, rebroadcasting domestic Russian TV shows and news throughout the Baltic States. This means that ethnic Russians throughout the region are exposed to the media formulated by state-controlled outlets in Russia. Web media is becoming increasingly popular among Northarian residents of all ethnicities, but ethnic Russians in the country report high levels of trust in Russian websites mail.ru and odnoklassniki.ru.

Polls and surveys conducted in Northaria indicate that ethnic Russians prefer the Russian state-controlled media outlets and trust them to provide better information than the Northarian or Western counterparts. This posture is problematic throughout the region because Russian media tends toward pointedly different narratives than Western media outlets, portraying NATO, the EU, and the U.S. as antagonists in most storylines and Russia as the protector against Western encroachment. The narrative running through most storylines in Russian-based media portrays the West as a bona fide threat to Russia and its diaspora in the Baltics, suggesting that the states' decision to join NATO and the EU has been an unqualified disaster for Russian populations there. Although the media does not necessarily make the claim that ethnic Russians in Northaria are less well off than Russians in Russia proper, the narrative does insist that ethnic discrimination, unequal conditions, and psychological pressure are rampant as ethnic Northarians continue to categorize ethnic Russians in their country as cultural invaders. The media frequently brings up the troubling specter of interethnic conflict should the Northarian government fail to address the issues appropriately. The language reforms in Northarian schools are trotted out with some regularity, posing the policy as an attempt to ensure that Russians known their second-class status in the country, not an honest attempt at assimilation and integration. Northarian nationalism, indeed any Baltic nationalism, is equated with Russophobia. Any efforts to remove or deface Soviet imagery are characterized as unrepentantly nationalist and barbaric, akin to inciting violent acts of revenge against Russian populations still remaining there. The perceived plight of ethnic Russians in Northaria is used by Russian propaganda as evidence that the Northarian government is fundamentally illegitimate due to its ethnocratic regime and suffering from a democracy deficit.

If the perspective remains unchallenged, it is likely that ethnic Russians in Northaria will continue to perceive the Northarian government, and its alliances with the West, as a threat. Northaria, alongside other Baltic States, has difficulties confronting Russia's information onslaught since its laws protect free speech. There is little constitutional wiggle room to restrict Russia's access to Northaria's information domain.

Analysis of Information and Infrastructure Dynamics in Northaria

a. The most conspicuous problem with national infrastructure is its deteriorating condition, particularly in remote and rural areas. To the degree that shortcomings reinforce ethnic divisions, the government should address the problem. Free public transportation might be replaced with modest fees in order to generate finances for infrastructure repair.

b. To address vulnerability to cyberwarfare, Northaria should take steps to strengthen defenses and reduce risk.

(1) Government functions and data should be backed up on servers that reside outside the country in secure locations. The government should undertake routine drills in the continuity of governance in the face of invasion.

(2) The government should conduct annual red-teaming drills that probe for vulnerability within critical infrastructure and continually improve defenses.

c. The government should consider providing incentives for national media to improve their performance and appeal to all residents of the country. Studies indicate that people will watch the highest quality programs, including for their news requirements. Northaria should consult with Western media companies to garner the necessary expertise to improve.

ENDNOTES

[1] The parties highlighted in blue are part of the governing coalition.

[2] USG STR Framework (Draft version .32), p. 23.

[3] Requested permission on 30 May 2016. http://en.academic.ru/pictures/enwiki/78/Nazi-Soviet_1941.png.

[4] Public domain
https://upload.wikimedia.org/wikipedia/commons/a/a0/Karlis_Ulmanis_1934.jpg.

[5] Requested permission 5 June 2016. Permission received 6 June 2016.
https://www.bing.com/images/search?q=map+of+latvia&view=detailv2&&id=066B3404F303BD6DA2C89039BB91CD8770DCFE75&selectedIndex=0&ccid=rCPtNtmS&simid=608043189751973244&thid=OIP.Mac23ed36d99288feebe722341bba62b0H0&ajaxhist=0

[6] JHU/APL, *Little Green Men*, 11.

BIBLIOGRAPHY

Bartkowski, Maciej. "Imaging Polish Nation: Nonviolent Resistance in Poland under Partitions," *Free Russia* (June 2015): 5.

Bleiere, Daina, Butulis, IIgvars, Feldmanis, Inesis, Stranga, Aivars, & Zunda, Antonijis. *History of Latvia: 100 Years* (Riga, Latvia: Society Domas Spēks, 2014), 122-23.

Foster, Peter & Holehouse, Matthew. "Russia Accused Of Clandestine Funding Of European Parties as U.S. Conducts Major Review of Vladimir Putin's Strategy," *The Telegraph* (18 July 2016). https://www.telegraph.co.uk/news/worldnews/europe/russia/12103602/America-to-investigate-Russian-meddling-in-EU.html.

Ganser, Daniele. *NATO's Secret Armies: Operation Gladio and Terrorism in Western Europe.* London and New York: Frank Cass, 2005.

Harding, Luke. "WikiLeaks Cables Condemn Russia As 'Mafia State'," *The Guardian.* (2010). https://www.theguardian.com/world/2010/dec/01/wikileaks-cables-russia-mafia-kleptocracy.

The Johns Hopkins University Applied Physics Laboratory (JHU/APL), *Little Green Men: A Primer on Modern Russian Unconventional Warfare, Ukraine 2013-2014.* Fort Bragg, North Carolina: United States Army Special Operations Command, 2015.

Kalniete, Sandra. *With Dance Shoes in Siberian Snow.* Riga: Latvijas Okupacijas Muzeja Biedriba, 2006.

Koshkin, *What are the Kremlin's New Red Lines in the Post-Soviet Space?* 2015.

Lanoszka, Alexander. "Russian Hybrid Warfare and Extended Deterrence in Eastern Europe," *International Affairs* 92, no. 1 (January 2016): 175-95.

Lipski, Jan Jozef. *KOR: A History of the Workers' Defense Committee in Poland, 1976-1981.* Berkeley: University of California Press, 1985.

McFaul, Michael. "Transitions from Postcommunism," *Journal of Democracy* 16, no. 3 (2005): 12.

McFaul, Michael. "Transitions from Postcommunism," *Journal of Democracy* 16, no. 3 (2005), 12.

Nakhoda, Zein. "Solidarnosc (Solidarity) Brings Down the Communist Government of Poland, 1988-89," *Global Nonviolent Action Database*,10 September 2011, http://nvdatabase.swarthmore.edu/content/solidarno-solidarity-brings-down-communist-government-poland-1988-89.

Nollendorfs, Valters, Celle, Ojārs, Michele, Gundega, Neiburgs, Uldis, & Stasš, Dagnija. *The Three Occupations of Latvia, 1940-1991: Soviet and NAZI Take-Overs and Their Consequences*. Riga: Latvia Occupation Museum, 2013.

Renz, Betina & Smith, Hannah. "Russia and Hybrid Warfare--Going Beyond the Label," Aleksanteri Papers, Aleksanteri Institute, University of Helsinki, Finland: Kikimora Publications (January 2016). www.helsinki.fi/aleksanteri/english/publications/aleksanteri_papers.html.

The Road to Gdansk, directed by Maxim Ford (1983, United Kingdom, Parallax Pictures), https://www.youtube. com/ watch?v=y00Fi48VI38, 17:00 – 18:00

USG STR Framework (Draft version .32), p. 23

Wheatley, Jonathan. *Georgia from National Awakening to Rose Revolution*. Burlington, VT: Ashgate, 2005.

9 781925 907513